FREE Test Taking Tips DVD Offer

SHSAT Prep Book

Specialized High School Admissions Study Guide With 3 New York City SHSAT Practice Tests for Math and ELA [6th Edition]

Joshua Rueda

Written and edited by TPB Publishing.

ISBN 13: 9781637751473
ISBN 10: 1637751478

Table of Contents

Quick Overview

As you draw closer to taking your exam, effective preparation becomes more and more important. Thankfully, you have this study guide to help you get ready. Use this guide to help keep your studying on track and refer to it often.

This study guide contains several key sections that will help you be successful on your exam. The guide contains tips for what you should do the night before and the day of the test. Also included are test-taking tips. Knowing the right information is not always enough. Many well-prepared test takers struggle with exams. These tips will help equip you to accurately read, assess, and answer test questions.

A large part of the guide is devoted to showing you what content to expect on the exam and to helping you better understand that content. In this guide are practice test questions so that you can see how well you have grasped the content. Then, answer explanations are provided so that you can understand why you missed certain questions.

Don't try to cram the night before you take your exam. This is not a wise strategy for a few reasons. First, your retention of the information will be low. Your time would be better used by reviewing information you already know rather than trying to learn a lot of new information. Second, you will likely become stressed as you try to gain a large amount of knowledge in a short amount of time. Third, you will be depriving yourself of sleep. So be sure to go to bed at a reasonable time the night before. Being well-rested helps you focus and remain calm.

Be sure to eat a substantial breakfast the morning of the exam. If you are taking the exam in the afternoon, be sure to have a good lunch as well. Being hungry is distracting and can make it difficult to focus. You have hopefully spent lots of time preparing for the exam. Don't let an empty stomach get in the way of success!

When travelling to the testing center, leave earlier than needed. That way, you have a buffer in case you experience any delays. This will help you remain calm and will keep you from missing your appointment time at the testing center.

Be sure to pace yourself during the exam. Don't try to rush through the exam. There is no need to risk performing poorly on the exam just so you can leave the testing center early. Allow yourself to use all of the allotted time if needed.

Remain positive while taking the exam even if you feel like you are performing poorly. Thinking about the content you should have mastered will not help you perform better on the exam.

Once the exam is complete, take some time to relax. Even if you feel that you need to take the exam again, you will be well served by some down time before you begin studying again. It's often easier to convince yourself to study if you know that it will come with a reward!

Test-Taking Strategies

1. Predicting the Answer

When you feel confident in your preparation for a multiple-choice test, try predicting the answer before reading the answer choices. This is especially useful on questions that test objective factual knowledge. By predicting the answer before reading the available choices, you eliminate the possibility that you will be distracted or led astray by an incorrect answer choice. You will feel more confident in your selection if you read the question, predict the answer, and then find your prediction among the answer choices. After using this strategy, be sure to still read all of the answer choices carefully and completely. If you feel unprepared, you should not attempt to predict the answers. This would be a waste of time and an opportunity for your mind to wander in the wrong direction.

2. Reading the Whole Question

Too often, test takers scan a multiple-choice question, recognize a few familiar words, and immediately jump to the answer choices. Test authors are aware of this common impatience, and they will sometimes prey upon it. For instance, a test author might subtly turn the question into a negative, or he or she might redirect the focus of the question right at the end. The only way to avoid falling into these traps is to read the entirety of the question carefully before reading the answer choices.

3. Looking for Wrong Answers

Long and complicated multiple-choice questions can be intimidating. One way to simplify a difficult multiple-choice question is to eliminate all of the answer choices that are clearly wrong. In most sets of answers, there will be at least one selection that can be dismissed right away. If the test is administered on paper, the test taker could draw a line through it to indicate that it may be ignored; otherwise, the test taker will have to perform this operation mentally or on scratch paper. In either case, once the obviously incorrect answers have been eliminated, the remaining choices may be considered. Sometimes identifying the clearly wrong answers will give the test taker some information about the correct answer. For instance, if one of the remaining answer choices is a direct opposite of one of the eliminated answer choices, it may well be the correct answer. The opposite of obviously wrong is obviously right! Of course, this is not always the case. Some answers are obviously incorrect simply because they are irrelevant to the question being asked. Still, identifying and eliminating some incorrect answer choices is a good way to simplify a multiple-choice question.

4. Don't Overanalyze

Anxious test takers often overanalyze questions. When you are nervous, your brain will often run wild, causing you to make associations and discover clues that don't actually exist. If you feel that this may be a problem for you, do whatever you can to slow down during the test. Try taking a deep breath or counting to ten. As you read and consider the question, restrict yourself to the particular words used by the author. Avoid thought tangents about what the author *really* meant, or what he or she was *trying* to say. The only things that matter on a multiple-choice test are the words that are actually in the question. You must avoid reading too much into a multiple-choice question, or supposing that the writer meant something other than what he or she wrote.

5. No Need for Panic

It is wise to learn as many strategies as possible before taking a multiple-choice test, but it is likely that you will come across a few questions for which you simply don't know the answer. In this situation, avoid panicking. Because most multiple-choice tests include dozens of questions, the relative value of a single wrong answer is small. As much as possible, you should compartmentalize each question on a multiple-choice test. In other words, you should not allow your feelings about one question to affect your success on the others. When you find a question that you either don't understand or don't know how to answer, just take a deep breath and do your best. Read the entire question slowly and carefully. Try rephrasing the question a couple of different ways. Then, read all of the answer choices carefully. After eliminating obviously wrong answers, make a selection and move on to the next question.

6. Confusing Answer Choices

When working on a difficult multiple-choice question, there may be a tendency to focus on the answer choices that are the easiest to understand. Many people, whether consciously or not, gravitate to the answer choices that require the least concentration, knowledge, and memory. This is a mistake. When you come across an answer choice that is confusing, you should give it extra attention. A question might be confusing because you do not know the subject matter to which it refers. If this is the case, don't eliminate the answer before you have affirmatively settled on another. When you come across an answer choice of this type, set it aside as you look at the remaining choices. If you can confidently assert that one of the other choices is correct, you can leave the confusing answer aside. Otherwise, you will need to take a moment to try to better understand the confusing answer choice. Rephrasing is one way to tease out the sense of a confusing answer choice.

7. Your First Instinct

Many people struggle with multiple-choice tests because they overthink the questions. If you have studied sufficiently for the test, you should be prepared to trust your first instinct once you have carefully and completely read the question and all of the answer choices. There is a great deal of research suggesting that the mind can come to the correct conclusion very quickly once it has obtained all of the relevant information. At times, it may seem to you as if your intuition is working faster even than your reasoning mind. This may in fact be true. The knowledge you obtain while studying may be retrieved from your subconscious before you have a chance to work out the associations that support it. Verify your instinct by working out the reasons that it should be trusted.

8. Key Words

Many test takers struggle with multiple-choice questions because they have poor reading comprehension skills. Quickly reading and understanding a multiple-choice question requires a mixture of skill and experience. To help with this, try jotting down a few key words and phrases on a piece of scrap paper. Doing this concentrates the process of reading and forces the mind to weigh the relative importance of the question's parts. In selecting words and phrases to write down, the test taker thinks about the question more deeply and carefully. This is especially true for multiple-choice questions that are preceded by a long prompt.

9. Subtle Negatives

One of the oldest tricks in the multiple-choice test writer's book is to subtly reverse the meaning of a question with a word like *not* or *except*. If you are not paying attention to each word in the question, you can easily be led astray by this trick. For instance, a common question format is, "Which of the following is...?" Obviously, if the question instead is, "Which of the following is not...?," then the answer will be quite different. Even worse, the test makers are aware of the potential for this mistake and will include one answer choice that would be correct if the question were not negated or reversed. A test taker who misses the reversal will find what he or she believes to be a correct answer and will be so confident that he or she will fail to reread the question and discover the original error. The only way to avoid this is to practice a wide variety of multiple-choice questions and to pay close attention to each and every word.

10. Reading Every Answer Choice

It may seem obvious, but you should always read every one of the answer choices! Too many test takers fall into the habit of scanning the question and assuming that they understand the question because they recognize a few key words. From there, they pick the first answer choice that answers the question they believe they have read. Test takers who read all of the answer choices might discover that one of the latter answer choices is actually *more* correct. Moreover, reading all of the answer choices can remind you of facts related to the question that can help you arrive at the correct answer. Sometimes, a misstatement or incorrect detail in one of the latter answer choices will trigger your memory of the subject and will enable you to find the right answer. Failing to read all of the answer choices is like not reading all of the items on a restaurant menu: you might miss out on the perfect choice.

11. Spot the Hedges

One of the keys to success on multiple-choice tests is paying close attention to every word. This is never truer than with words like almost, most, some, and sometimes. These words are called "hedges" because they indicate that a statement is not totally true or not true in every place and time. An absolute statement will contain no hedges, but in many subjects, the answers are not always straightforward or absolute. There are always exceptions to the rules in these subjects. For this reason, you should favor those multiple-choice questions that contain hedging language. The presence of qualifying words indicates that the author is taking special care with his or her words, which is certainly important when composing the right answer. After all, there are many ways to be wrong, but there is only one way to be right! For this reason, it is wise to avoid answers that are absolute when taking a multiple-choice test. An absolute answer is one that says things are either all one way or all another. They often include words like *every*, *always*, *best*, and *never*. If you are taking a multiple-choice test in a subject that doesn't lend itself to absolute answers, be on your guard if you see any of these words.

12. Long Answers

In many subject areas, the answers are not simple. As already mentioned, the right answer often requires hedges. Another common feature of the answers to a complex or subjective question are qualifying clauses, which are groups of words that subtly modify the meaning of the sentence. If the question or answer choice describes a rule to which there are exceptions or the subject matter is complicated, ambiguous, or confusing, the correct answer will require many words in order to be expressed clearly and accurately. In essence, you should not be deterred by answer choices that seem excessively long. Oftentimes, the author of the text will not be able to write the correct answer without offering some qualifications and modifications. Your job is to read the answer choices thoroughly and completely and to select the one that most accurately and precisely answers the question.

13. Restating to Understand

Sometimes, a question on a multiple-choice test is difficult not because of what it asks but because of how it is written. If this is the case, restate the question or answer choice in different words. This process serves a couple of important purposes. First, it forces you to concentrate on the core of the question. In order to rephrase the question accurately, you have to understand it well. Rephrasing the question will concentrate your mind on the key words and ideas. Second, it will present the information to your mind in a fresh way. This process may trigger your memory and render some useful scrap of information picked up while studying.

14. True Statements

Sometimes an answer choice will be true in itself, but it does not answer the question. This is one of the main reasons why it is essential to read the question carefully and completely before proceeding to the answer choices. Too often, test takers skip ahead to the answer choices and look for true statements. Having found one of these, they are content to select it without reference to the question above. Obviously, this provides an easy way for test makers to play tricks. The savvy test taker will always read the entire question before turning to the answer choices. Then, having settled on a correct answer choice, he or she will refer to the original question and ensure that the selected answer is relevant. The mistake of choosing a correct-but-irrelevant answer choice is especially common on questions related to specific pieces of objective knowledge. A prepared test taker will have a wealth of factual knowledge at his or her disposal, and should not be careless in its application.

15. No Patterns

One of the more dangerous ideas that circulates about multiple-choice tests is that the correct answers tend to fall into patterns. These erroneous ideas range from a belief that B and C are the most common right answers, to the idea that an unprepared test-taker should answer "A-B-A-C-A-D-A-B-A." It cannot be emphasized enough that pattern-seeking of this type is exactly the WRONG way to approach a multiple-choice test. To begin with, it is highly unlikely that the test maker will plot the correct answers according to some predetermined pattern. The questions are scrambled and delivered in a random order. Furthermore, even if the test maker was following a pattern in the assignation of correct answers, there is no reason why the test taker would know which pattern he or she was using. Any attempt to discern a pattern in the answer choices is a waste of time and a distraction from the real work of taking the test. A test taker would be much better served by extra preparation before the test than by reliance on a pattern in the answers.

FREE DVD OFFER

Don't forget that doing well on your exam includes both understanding the test content and understanding how to use what you know to do well on the test. We offer a completely FREE Test Taking Tips DVD that covers world class test taking tips that you can use to be even more successful when you are taking your test.

All that we ask is that you email us your feedback about your study guide. To get your **FREE Test Taking Tips DVD**, email freedvd@studyguideteam.com with "FREE DVD" in the subject line and the following information in the body of the email:

- The title of your study guide.
- Your product rating on a scale of 1-5, with 5 being the highest rating.
- Your feedback about the study guide. What did you think of it?
- Your full name and shipping address to send your free DVD.

Introduction to the SHSAT

Function of the Test

The SHSAT is a standardized test that is used as the sole factor for admission to eight of New York City's Specialized High Schools. Fiorello H. LaGuardia High School is the only Specialized High School in New York City that does not require students to take the SHSAT exam as part of the admissions process. Students who are in the eighth or ninth grades who wish to attend one of these eight Specialized High Schools and who live in the five boroughs of New York City (Brooklyn, Manhattan, Queens, Staten Island, and The Bronx) must sit for this exam.

Test Administration

Each year, the SHSAT test is only offered in the month of October for eighth grade students and in the month of November for ninth grade students. Students who are interested in registering to take the exam can do so by talking with their school's guidance counselor. After students are registered, they will receive a test ticket to sit for the exam. Students who are sitting for the SHSAT exam must also rank (in order of priority) the Specialized High Schools that they would like to attend on their test ticket.

Students are able to take the SHSAT test twice—once in the eighth grade and once in the ninth grade—if they are not accepted to the Specialized High School of their choice after taking the exam in the eighth grade.

Students will be provided with the necessary accommodations for taking the exam, as long as the accommodations are permitted for the test. If necessary, mathematics glossaries can be provided in nine languages to students on the day of the exam.

Test Format

Students are given 180 minutes to complete the SHSAT, which is comprised of 57 questions in each of its two sections: English language arts (ELA) and math, as outlined in the table. All of the reading and writing questions in the ELA section are multiple-choice and split between two categories. The first category requires students to utilize their revising and editing skills, while the second category assesses reading comprehension by asking students to extract information from various reading passages in order to answer associated questions. In the math section, there are 52 multiple-choice questions that deal with word and computational problems, as well as five grid-in questions that are *not* multiple-choice. These questions require students to provide correct numerical solutions to computational problems. Finally, all multiple-choice questions on both sections of the test have four answer choices per question, and both sections of the test each have ten unscored experimental questions that are used for field testing purposes for future iterations of the exam.

Sections of the SHSAT Test			
Subject Areas	**Questions (Multiple-Choice)***	**Question breakdown**	**Time Limit**
English Language Arts (ELA) Reading & Writing	57	9-11 revising/editing	180 minutes
		46-48 reading comprehension questions 3-4 informational passages 1-2 literary prose passages 1 poem	
		10 embedded field questions (unscored)	
Math	57	52 word & computational problems	
		5 grid-in questions*	
		10 embedded field questions (unscored)	
Total Questions:	114		

**Students are required to provide correct numerical answers for these questions*

Scoring

Individuals are not penalized for wrong answers or for questions that are left blank. After completing the test, each student is given a raw score that is based on the number of questions answered correctly. Those raw scores are then converted into three-digit composite scores (an 800 being the highest possible score). Scores are made available to the schools in March following the fall in which the exam was taken. For example, if a student takes the exam in the fall of 2020, his or her score will be released to the schools in the March of 2021.

Once the test results are in, all students who took the SHSAT exam are ranked in order by composite score from highest to lowest. Seats are then filled in each of the Specialized High Schools, in order, according to the students' first choices until all of the open seats for that academic year are filled. The number of available seats at each of the Specialized High Schools varies from year to year.

Recent/Future Developments

The SHSAT was revised in 2018 with changes mainly to the English language arts (ELA) section. The amount of questions in the Revising/Editing section decreased from 20 to 9-11. The Reading Comprehension section previously included primarily informational texts to analyze. The update now includes informational passages as well as literary prose passages and one poem on which questions are based.

Study Prep Plan for the SHSAT

1 **Schedule -** Use one of our study schedules below or come up with one of your own.

2 **Relax -** Test anxiety can hurt even the best students. There are many ways to reduce stress. Find the one that works best for you.

3 **Execute -** Once you have a good plan in place, be sure to stick to it.

One Week Study Schedule

Day 1	Reading Comprehension
Day 2	Revising/Editing
Day 3	Practice Questions
Day 4	Numbers and Operations
Day 5	Probability and Statistics
Day 6	Practice Questions
Day 7	Take Your Exam!

Two Week Study Schedule

Day 1	Reading Comprehension	Day 8	Algebra
Day 2	Understanding the Effect of Word Choice	Day 9	Solving Systems of Equations
Day 3	Revising/Editing	Day 10	Probability and Statistics
Day 4	Word Usage	Day 11	Geometry and Measurements
Day 5	Practice Questions	Day 12	Practice Questions
Day 6	Answer Explanations	Day 13	Answer Explanations
Day 7	Numbers and Operations	Day 14	Take Your Exam!

One Month Study Schedule

Day 1	Main Ideas and Supporting Details	Day 11	Forming Paragraphs	Day 21	Representing and Solving Equations and Inequalities Graphically
Day 2	Understanding the Development of Themes	Day 12	Practice Questions	Day 22	Function and Function Notation
Day 3	Understanding Poetic Devices and Structure	Day 13	Answer Explanations	Day 23	Probability and Statistics
Day 4	Understanding the Characteristics of Literary Genres	Day 14	Properties of Operations with Real Numbers	Day 24	Representing Data
Day 5	Identifying Primary Sources in Media	Day 15	Properties of Rational and Irrational Numbers	Day 25	Geometry and Measurements
Day 6	Parts of Speech	Day 16	Choosing a Level of Accuracy	Day 26	Perimeter and Area
Day 7	Verbs	Day 17	Algebra	Day 27	Transformations in the Plane
Day 8	Subject-Verb Agreement	Day 18	Writing Expressions in Equivalent Forms	Day 28	Practice Questions
Day 9	Word Parts	Day 19	Solving Equations	Day 29	Answer Explanations
Day 10	Sentence Fluency	Day 20	The Quadratic Formula	Day 30	Take Your Exam!

English Language Arts

Reading Comprehension

Main Ideas and Supporting Details

Topics and main ideas are critical parts of writing. The **topic** is the subject matter of the piece. An example of a topic would be global warming.

The **main idea** is what the writer wants to say about that topic. A writer may make the point that global warming is a growing problem that must be addressed in order to save the planet. Therefore, the topic is global warming, and the main idea is that it's a serious problem needing to be addressed. The topic can be expressed in a word or two, but the main idea should be a complete thought.

An author will likely identify the topic immediately within the title or the first sentence of a passage. The main idea is usually presented in the introduction. In a single passage, the main idea may be identified in the first or last sentence, but it will most likely be directly stated and easily recognized by the reader. Because it is not always stated immediately in a passage, it's important to carefully read the entire passage to identify the main idea.

The main idea should not be confused with the thesis statement. A **thesis statement** is a clear statement of the writer's specific stance and can often be found in the introduction of a non-fiction piece. The thesis is a specific sentence (or two) that offers the direction and focus of the discussion.

In order to illustrate the main idea, a writer will use **supporting details**, the details that provide evidence or examples to help make a point. Supporting details often appear in the form of quotations, paraphrasing, or analysis. Authors should connect details and analysis to the main point.

For example, in the example of global warming, where the author's main idea is to show the seriousness of this growing problem and the need for change, the use of supporting details in this piece would be critical in effectively making that point. Supporting details used here might include statistics on an increase in global temperatures and studies showing the impact of global warming on the planet. The author could also include projections for future climate change in order to illustrate potential lasting effects of global warming.

It's important to evaluate the author's supporting details to be sure that they are credible, provide evidence of the author's point, and directly support the main idea. Though shocking statistics grab readers' attention, their use could be ineffective information in the piece. Details like this are crucial to understanding the passage and evaluating how well the author presents his or her argument and evidence.

Also remember that when most authors write, they want to make a point or send a message. This point or message of a text is known as the theme. Authors may state themes explicitly, like in *Aesop's Fables*. More often, especially in modern literature, readers must infer the theme based on text details. Usually after carefully reading and analyzing an entire text, the reader can identify the theme. Typically, the longer the piece, the more themes you will encounter, though often one theme dominates the rest, as evidenced by the author's purposeful revisiting of it throughout the passage.

Summarizing a Complex Text

A **summary** is a shortened version of the original text, written by the reader in their own words. In order to effectively summarize a more complex text, it is necessary to fully understand the original source, and to highlight the major points covered. It may be helpful to outline the original text to get a big picture view of it, and to avoid getting bogged down in the minor details. For example, a summary wouldn't need to include a specific statistic from the original source unless it was the major focus of the piece. Also, it's important for readers to use their own words but to retain the original meaning of the passage. The key to a good summary is to emphasize the main idea without changing the focus of the original information.

Paraphrasing calls for the reader to take a small part of the passage and list or describe its main points. Paraphrasing is more than rewording the original passage, though. As with summary, a paraphrase should be written in the reader's own words, while still retaining the meaning of the original source. The main difference between summarizing and paraphrasing is the length of the original passage. A summary would be appropriate for a much larger piece, while paraphrase might focus on just a few lines of text. Effective paraphrasing will indicate an understanding of the original source, yet still help the reader expand on their interpretation. A paraphrase should neither add new information nor remove essential facts that will change the meaning of the source.

Recognizing the Structure of Texts

Writing can be classified under four passage types: narrative, expository, descriptive (sometimes called technical), and persuasive. Though these types are not mutually exclusive, one form tends to dominate the rest. By recognizing the *type* of passage you're reading, you gain insight into *how* you should read. If you're reading a narrative, you can assume the author intends to entertain, which means you may skim the text without losing meaning. A technical document might require a close read because skimming the passage might cause the reader to miss salient details.

1. **Narrative writing**, at its core, is the art of storytelling. For a narrative to exist, certain elements must be present. First, it must have characters. While many characters are human, characters could be defined as anything that thinks, acts, and talks like a human. For example, many recent movies, such as *Lord of the Rings* and *The Chronicles of Narnia*, include animals, fantastical creatures, and even trees that behave like humans. Second, it must have a plot or sequence of events. Typically, those events follow a standard plot diagram, but recent trends start *in medias res* or in the middle (near the climax). In this instance, foreshadowing and flashbacks often fill in plot details. Finally, along with characters and a plot, there must also be conflict. Conflict is usually divided into two types: internal and external. Internal conflict indicates the character is in turmoil and is presented through the character's thoughts. External conflicts are visible. Types of external conflict include a person versus nature, another person, or society.

2. **Expository writing** *is detached and to the point. Since expository* writing is designed to instruct or inform, it usually involves directions and steps written in second person ("you" voice) and lacks any persuasive or narrative elements. Sequence words such as *first*, *second*, and *third*, or *in the first place*, *secondly*, and *lastly* are often given to add fluency and cohesion. Common examples of expository writing include instructor's lessons, cookbook recipes, and repair manuals.

3. Due to its empirical nature, **technical writing** is filled with steps, charts, graphs, data, and statistics. The goal of technical writing is to advance understanding in a field through the scientific method. Experts such as teachers, doctors, or mechanics use words unique to the profession in which they

operate. These words, which often incorporate acronyms, are called *jargon*. Technical writing is a type of expository writing but is not meant to be understood by the general public. Instead, technical writers assume readers have received a formal education in a particular field of study and need no explanation as to what the jargon means. Imagine a doctor trying to understand a diagnostic reading for a car or a mechanic trying to interpret lab results. Only professionals with proper training will fully comprehend the text.

4. **Persuasive writing** is designed to change opinions and attitudes. The topic, stance, and arguments are found in the thesis, positioned near the end of the introduction. Later supporting paragraphs offer relevant quotations, paraphrases, and summaries from primary or secondary sources, which are then interpreted, analyzed, and evaluated. The goal of persuasive writers is not to stack quotes but to develop original ideas by using sources as a starting point. Good persuasive writing makes powerful arguments with valid sources and thoughtful analysis. Poor persuasive writing is riddled with bias and logical fallacies. Sometimes logical and illogical arguments are sandwiched together in the same piece. Therefore, readers should display skepticism when reading persuasive arguments.

Understanding Literary Interpretation

Literary interpretation is an interpretation and analysis of a literary work, based on the textual evidence in the work. It is often subjective as critical readers may discern different meanings in the details. A test taker needs to be prepared for questions that will test how well he or she can read a passage, make an analysis, and then provide evidence to support that analysis.

Literal and Figurative Meanings

When analyzing and interpreting fiction, readers must be active participants in the experience. Some authors make their messages clearer than others, but the onus is on the reader to add layers to what is read through interpretation. In literary interpretation, the goal is not to offer an opinion as to the inherent value of the work. Rather, the goal is to determine what the text means by analyzing the literal and figurative meanings of the text through critical reading.

Critical reading is close reading that elicits questions as the reader progresses. Many authors of fiction use literary elements and devices to further theme and to speak to their audience. These elements often utilize language that has an alternate or figurative meaning in addition to their actual or literal meaning. Readers should be asking questions about these and other important details as a passage is analyzed. What unfamiliar words are there? What is their contextual definition? How do they contribute to the overall feel of the work? How do they contribute to the mood and general message?

Drawing Inferences

An **inference** refers to a point that is implied (as opposed to directly-stated) by the evidence presented. It's necessary to use inference in order to draw conclusions about the meaning of a passage. Authors make implications through character dialogue, thoughts, effects on others, actions, and looks.

When making an inference about a passage, it's important to rely only on the information that is provided in the text itself. This helps readers ensure that their conclusions are valid.

Textual Evidence

It's helpful to read a passage a few times, noting details that seem important to the piece. Textual evidence helps readers draw a conclusion about a passage. **Textual evidence** refers to information—facts and examples that support the main idea. Textual evidence will likely come from outside sources and can be in the form of quoted or paraphrased material. In order to draw a conclusion from evidence, it's important to examine the credibility and validity of that evidence as well as how (and if) it relates to the main idea. Effective use of textual evidence should connect to the main idea and support a specific point.

Understanding the Development of Themes

Identifying Theme or Central Message

The **theme** is the central message of a fictional work, whether that work is structured as prose, drama, or poetry. It is the heart of what an author is trying to say to readers through the writing, and theme is largely conveyed through literary elements and techniques.

In literature, a theme can often be determined by considering the over-arching narrative conflict within the work. Though there are several types of conflicts and several potential themes within them, the following are the most common:

- *Individual against the self*—relevant to themes of self-awareness, internal struggles, pride, coming of age, facing reality, fate, free will, vanity, loss of innocence, loneliness, isolation, fulfillment, failure, and disillusionment
- *Individual against nature*— relevant to themes of knowledge vs. ignorance, nature as beauty, quest for discovery, self-preservation, chaos and order, circle of life, death, and destruction of beauty
- *Individual against society*— relevant to themes of power, beauty, good, evil, war, class struggle, totalitarianism, role of men/women, wealth, corruption, change vs. tradition, capitalism, destruction, heroism, injustice, and racism
- *Individual against another individual*— relevant to themes of hope, loss of love or hope, sacrifice, power, revenge, betrayal, and honor

For example, in Hawthorne's *The Scarlet Letter*, one possible narrative conflict could be the individual against the self, with a relevant theme of internal struggles. This theme is alluded to through characterization—Dimmesdale's moral struggle with his love for Hester and Hester's internal struggles with the truth and her daughter, Pearl. It's also alluded to through plot—Dimmesdale's suicide and Hester helping the very townspeople who initially condemned her.

Sometimes, a text can convey a message or universal lesson—a truth or insight that the reader infers from the text, based on analysis of the literary and/or poetic elements. This message is often presented as a statement. For example, a potential message in Shakespeare's *Hamlet* could be "Revenge is what ultimately drives the human soul." This message can be immediately determined through plot and characterization in numerous ways, but it can also be determined through the setting of Norway, which is bordering on war.

How Authors Develop Theme

Authors employ a variety of techniques to present a theme. They may compare or contrast characters, events, places, ideas, or historical or invented settings to speak thematically. They may use analogies, metaphors, similes, allusions, or other literary devices to convey the theme. An author's use of diction, syntax, and tone can also help convey the theme. Authors will often develop themes through the development of characters, use of the setting, repetition of ideas, use of symbols, and through contrasting value systems. Authors of both fiction and nonfiction genres will use a variety of these techniques to develop one or more themes.

Regardless of the literary genre, there are commonalities in how authors, playwrights, and poets develop themes or central ideas.

Authors often do research, the results of which contributes to theme. In prose fiction and drama, this research may include real historical information about the setting the author has chosen or include elements that make fictional characters, settings, and plots seem realistic to the reader. In nonfiction, research is critical since the information contained within this literature must be accurate and, moreover, accurately represented.

In fiction, authors present a narrative conflict that will contribute to the overall theme. In fiction, this conflict may involve the storyline itself and some trouble within characters that needs resolution. In nonfiction, this conflict may be an explanation or commentary on factual people and events.

Authors will sometimes use character motivation to convey theme, such as in the example from *Hamlet* regarding revenge. In fiction, the characters an author creates will think, speak, and act in ways that effectively convey the theme to readers. In nonfiction, the characters are factual, as in a biography, but authors pay particular attention to presenting those motivations to make them clear to readers.

Authors also use literary devices as a means of conveying theme. For example, the use of moon symbolism in Mary Shelley's *Frankenstein* is significant as its phases can be compared to the phases that the Creature undergoes as he struggles with his identity.

The selected point of view can also contribute to a work's theme. The use of first-person point of view in a fiction or non-fiction work engages the reader's response differently than third person point of view. The central idea or theme from a first-person narrative may differ from a third-person limited text.

In literary nonfiction, authors usually identify the purpose of their writing, which differs from fiction, where the general purpose is to entertain. The purpose of nonfiction is usually to inform, persuade, or entertain the audience. The stated purpose of a non-fiction text will drive how the central message or theme, if applicable, is presented.

Authors identify an audience for their writing, which is critical in shaping the theme of the work. For example, the audience for J.K. Rowling's *Harry Potter* series would be different than the audience for a biography of George Washington. The audience an author chooses to address is closely tied to the purpose of the work. The choice of an audience also drives the choice of language and level of diction an author uses. Ultimately, the intended audience determines the level to which that subject matter is presented and the complexity of the theme.

Identifying Literary Elements

There is no one, final definition of what literary elements are. They can be considered features or characteristics of fiction, but they are really more of a way that readers can unpack a text for the purpose of analysis and understanding the meaning. The elements contribute to a reader's literary interpretation of a passage as to how they function to convey the central message of a work. The most common literary elements used for analysis are presented below.

Point of View

The **point of view** is the position the narrator takes when telling the story in prose. If a narrator is incorporated in a drama, the point of view may vary; in poetry, point of view refers to the position the speaker in a poem takes.

First Person

The first person point of view is when the writer uses the word "I" in the text. Poetry often uses first person, e.g., William Wordsworth's "I Wandered Lonely as a Cloud." Two examples of prose written in first person are Suzanne Collins' *The Hunger Games* and Anthony Burgess's *A Clockwork Orange*.

Second Person

The second person point of view is when the writer uses the pronoun "you." It is not widely used in prose fiction, but as a technique, it has been used by writers such as William Faulkner in *Absalom, Absalom!* and Albert Camus in *The Fall*. It is more common in poetry—e.g., Pablo Neruda's "If You Forget Me."

Third Person

Third person point of view is when the writer utilizes pronouns such as him, her, or them. It may be the most utilized point of view in prose as it provides flexibility to an author and is the one with which readers are most familiar. There are two main types of third person used in fiction. **Third person omniscient** uses a narrator that is all-knowing, relating the story by conveying and interpreting thoughts/feelings of all characters. In **third person limited**, the narrator relates the story through the perspective of one character's thoughts/feelings, usually the main character.

Plot

The **plot** is what happens in the story. Plots may be singular, containing one problem, or they may be very complex, with many sub-plots. All plots have exposition, a conflict, a climax, and a resolution. The **conflict** drives the plot and is something that the reader expects to be resolved. The plot carries those events along until there is a resolution to the conflict.

Tone

The **tone** of a story reflects the author's attitude and opinion about the subject matter of the story or text. Tone can be expressed through word choice, imagery, figurative language, syntax, and other details. The emotion or mood the reader experiences relates back to the tone of the story. Some examples of possible tones are humorous, somber, sentimental, and ironic.

Setting

The **setting** is the time, place, or set of surroundings in which the story occurs. It includes time or time span, place(s), climates, geography—man-made or natural—or cultural environments. Emily Dickinson's poem "Because I could not stop for Death" has a simple setting—the narrator's symbolic ride with Death through town towards the local graveyard. Conversely, Leo Tolstoy's *War and Peace* encompasses numerous settings within settings in the areas affected by the Napoleonic Wars, spanning 1805 to 1812.

Characters

Characters are the story's figures that assume primary, secondary, or minor roles. Central or major characters are those integral to the story—the plot cannot be resolved without them. A central character can be a **protagonist** or hero. There may be more than one protagonist, and he/she doesn't always have to possess good characteristics. A character can also be an **antagonist**—the force against a protagonist.

Character development is when the author takes the time to create dynamic characters that add uniqueness and depth to the story. *Dynamic* characters are characters that change over the course of the plot time. Stock characters are those that appear across genres and embrace stereotypes—e.g., the cowboy of the Wild West or the blonde bombshell in a detective novel. A flat character is one that does not present a lot of complexity or depth, while a rounded character does. Sometimes, the narrator of a story or the speaker in a poem can be a character—e.g., Nick Carraway in F. Scott Fitzgerald's *The Great Gatsby* or the speaker in Robert Browning's "My Last Duchess." The narrator might also function as a character in prose, though not be part of the story—e.g., Charles Dickens' narrator of *A Christmas Carol*.

Figurative Language

Whereas literal language is the author's use of precise words, proper meanings, definitions, and phrases that mean exactly what they say, **figurative language** deviates from precise meaning and word definition—often in conjunction with other familiar words and phrases—to paint a picture for the reader. Figurative language is less explicit and more open to reader interpretation.

Some examples of figurative language are included in the following graphic.

	Definition	**Example**
Simile	Compares two things using "like" or "as"	Her hair was like gold.
Metaphor	Compares two things as if they are the same	He was a giant teddy bear.
Idiom	Using words with predictable meanings to create a phrase with a different meaning	The world is your oyster.
Alliteration	Repeating the same beginning sound or letter in a phrase for emphasis	The busy baby babbled.
Personification	Attributing human characteristics to an object or an animal	The house glowered menacingly with a dark smile.
Foreshadowing	Giving an indication that something is going to happen later in the story	I wasn't aware at the time, but I would come to regret those words.
Symbolism	Using symbols to represent ideas and provide a different meaning	The ring represented the bond between us.
Onomatopoeia	Using words that imitate sound	The tire went off with a bang and a crunch.
Imagery	Appealing to the senses by using descriptive language	The sky was painted with red and pink and streaked with orange.
Hyperbole	Using exaggeration not meant to be taken literally	The girl weighed less than a feather.

Figurative language can be used to give additional insight into the theme or message of a text by moving beyond the usual and literal meaning of words and phrases. It can also be used to appeal to the senses of readers and create a more in-depth story.

Understanding Poetic Devices and Structure

Poetic Devices

Rhyme is the poet's use of corresponding word sounds in order to create an effect. Most rhyme occurs at the ends of a poem's lines, which is how readers arrive at the rhyme scheme. Each line that has a corresponding rhyming sound is assigned a letter—A, B, C, and so on. When using a rhyme scheme, poets will often follow lettered patterns. Robert Frost's *"The Road Not Taken"* uses the ABAAB rhyme scheme:

> Two roads diverged in a yellow wood, A
> And sorry I could not travel both B
> And be one traveler, long I stood A
> And looked down one as far as I could A
> To where it bent in the undergrowth; B

Another important poetic device is **rhythm**—metered patterns within poetry verses. When a poet develops rhythm through meter, he or she is using a combination of stressed and unstressed syllables to create a sound effect for the reader.

Rhythm is created by the use of poetic feet—individual rhythmic units made up of the combination of stressed and unstressed syllables. A line of poetry is made up of one or more poetic feet. There are five standard types in English poetry, as depicted in the chart below.

Foot Type	Rhythm	Pattern
Iamb	buh Buh	Unstressed/stressed
Trochee	Buh buh	Stressed/unstressed
Spondee	Buh Buh	Stressed/stressed
Anapest	buh buh Buh	Unstressed/unstressed/stressed
Dactyl	Buh buh buh	Stressed/unstressed/unstressed

Structure

Poetry is most easily recognized by its structure, which varies greatly. For example, a structure may be strict in the number of lines it uses. It may use rhyming patterns or may not rhyme at all. There are three main types of poetic structures:

- *Verse*—poetry with a consistent meter and rhyme scheme
- *Blank verse*—poetry with consistent meter but an inconsistent rhyme scheme
- *Free verse*—poetry with inconsistent meter or rhyme

Verse poetry is most often developed in the form of stanzas—groups of word lines. Stanzas can also be considered verses. The structure is usually formulaic and adheres to the protocols for the form. For example, the English sonnet form uses a structure of fourteen lines and a variety of different rhyming patterns. The English ode typically uses three ten-line stanzas and has a particular rhyming pattern.

Poets choose poetic structure based on the effect they want to create. Some structures—such as the ballad and haiku—developed out of cultural influences and common artistic practice in history, but in more modern poetry, authors choose their structure to best fit their intended effect.

Understanding the Effect of Word Choice

An author's choice of words—also referred to as **diction**—helps to convey his or her meaning in a particular way. Through diction, an author can convey a particular tone—e.g., a humorous tone, a serious tone—in order to support the thesis in a meaningful way to the reader.

Connotation and Denotation

Connotation is when an author chooses words or phrases that invoke ideas or feelings other than their literal meaning. An example of the use of connotation is the word *cheap*, which suggests something is poor in value or negatively describes a person as reluctant to spend money. When something or someone is described this way, the reader is more inclined to have a particular image or feeling about it or him/her. Thus, connotation can be a very effective language tool in creating emotion and swaying opinion. However, connotations are sometimes hard to pin down because varying emotions can be associated with a word. Generally, though, connotative meanings tend to be fairly consistent within a specific cultural group.

Denotation refers to words or phrases that mean exactly what they say. It is helpful when a writer wants to present hard facts or vocabulary terms with which readers may be unfamiliar. Some examples of denotation are the words inexpensive and frugal. *Inexpensive* refers to the cost of something, not its value, and *frugal* indicates that a person is conscientiously watching his or her spending. These terms do not elicit the same emotions that *cheap* does.

Authors sometimes choose to use both, but what they choose and when they use it is what critical readers need to differentiate. One method isn't inherently better than the other; however, one may create a better effect, depending upon an author's intent. If, for example, an author's purpose is to inform, to instruct, and to familiarize readers with a difficult subject, his or her use of connotation may be helpful. However, it may also undermine credibility and confuse readers. An author who wants to create a credible, scholarly effect in his or her text would most likely use denotation, which emphasizes literal, factual meaning and examples.

Technical Language

Test takers and critical readers alike should be very aware of technical language used within informational text. **Technical language** refers to terminology that is specific to a particular industry and is best understood by those specializing in that industry. This language is fairly easy to differentiate, since it will most likely be unfamiliar to readers. It's critical to be able to define technical language either by the author's written definition, through the use of an included glossary—if offered—or through context clues that help readers clarify word meaning.

Identifying Rhetorical Strategies

Rhetoric refers to an author's use of particular strategies, appeals, and devices to persuade an intended audience. The more effective the use of rhetoric, the more likely the audience will be persuaded.

Determining an Author's Point of View

A **rhetorical strategy**—also referred to as a rhetorical mode—is the structural way an author chooses to present his/her argument.

The following are some of the more common rhetorical strategies:

- *Cause and effect*—establishing a logical correlation or causation between two ideas
- *Classification/division*—the grouping of similar items together or division of something into parts
- *Comparison/contrast*—the distinguishing of similarities/differences to expand on an idea
- *Definition*—used to clarify abstract ideas, unfamiliar concepts, or to distinguish one idea from another
- *Description*—use of vivid imagery, active verbs, and clear adjectives to explain ideas
- *Exemplification*—the use of examples to explain an idea
- *Narration*—anecdotes or personal experience to present or expand on a concept
- *Problem/Solution*—presentation of a problem or problems, followed by proposed solution(s)

Rhetorical Strategies and Devices

A **rhetorical device** is the phrasing and presentation of an idea that reinforces and emphasizes a point in an argument. A rhetorical device is often quite memorable. One of the more famous uses of a rhetorical device is in John F. Kennedy's 1961 inaugural address: "Ask not what your country can do for you, ask what you can do for your country." The contrast of ideas presented in the phrasing is an example of the rhetorical device of antimetabole.

Some other common examples are provided below, but test takers should be aware that this is not a complete list.

Device	Definition	Example
Allusion	A reference to a famous person, event, or significant literary text as a form of significant comparison	"We are apt to shut our eyes against a painful truth, and listen to the song of that siren till she transforms us into beasts." Patrick Henry
Anaphora	The repetition of the same words at the beginning of successive words, phrases, or clauses, designed to emphasize an idea	"We shall not flag or fail. We shall go on to the end. We shall fight in France, we shall fight on the seas and oceans, we shall fight with growing confidence … we shall fight in the fields and in the streets, we shall fight in the hills. We shall never surrender." Winston Churchill
Understatement	A statement meant to portray a situation as less important than it actually is to create an ironic effect	"The war in the Pacific has not necessarily developed in Japan's favor." Emperor Hirohito, surrendering Japan in World War II
Parallelism	A syntactical similarity in a structure or series of structures used for impact of an idea, making it memorable	"A penny saved is a penny earned." Ben Franklin
Rhetorical question	A question posed that is not answered by the writer though there is a desired response, most often designed to emphasize a point	"Can anyone look at our reduced standing in the world today and say, 'Let's have four more years of this?'" Ronald Reagan

Literature refers to a collection of written works that are the distinctive voices of peoples, time periods, and cultures. The world has gained great insight into human thought, vices, virtues, and desires through the written word. As the work pertains to the author's approach to these insights, literature can be classified as fiction or non-fiction.

Understanding the Characteristics of Literary Genres

Classifying literature involves an understanding of the concept of genre. A *genre* is a category of literature that possesses similarities in style and in characteristics. Based on form and structure, there are four basic genres.

Fictional Prose

Fictional prose consists of fictional works written in standard form with a natural flow of speech and without poetic structure. Fictional prose primarily utilizes grammatically complete sentences and a paragraph structure to convey its message.

Drama

Drama is fiction that is written to be performed in a variety of media, intended to be performed for an audience, and structured for that purpose. It might be composed using poetry or prose, often straddling the elements of both in what actors are expected to present. Action and dialogue are the tools used in drama to tell the story.

Poetry

Poetry is fiction in verse that has a unique focus on the rhythm of language and focuses on intensity of feeling. It is not an entire story, though it may tell one; it is compact in form and in function. Poetry can be considered as a poet's brief word picture for a reader. Poetic structure is primarily composed of lines and stanzas. Together, poetic structure and devices are the methods that poets use to lead readers to feeling an effect and, ultimately, to the interpretive message.

Literary Nonfiction

Literary nonfiction is prose writing that is based on current or past real events or real people and includes straightforward accounts as well as those that offer opinions on facts or factual events. It is helpful to distinguish between literary nonfiction—a form of writing that incorporates literary styles and techniques to create factually-based narratives—and informational texts, which will be addressed in the next section.

Informational Texts

Informational texts are a category of texts within the genre of nonfiction. Their intent is to inform, and while they do convey a point of view and may include literary devices, they do not utilize other literary elements, such as characters or plot. An informational text also reflects a thesis—an implicit or explicit statement of the text's intent and/or a main idea—the overarching focus and/or purpose of the text, generally implied. Some examples of informational texts are informative articles, instructional/how-to texts, factual reports, reference texts, and self-help texts.

Interpreting Textual Evidence in Informational Text

Literal and Figurative Meanings

It is important when evaluating informational texts to consider the use of both literal and figurative meanings. The words and phrases an author chooses to include in a text must be evaluated. How does the word choice affect the meaning and tone? By recognizing the use of literal and figurative language, a reader can more readily ascertain the message or purpose of a text. Literal word choice is the easiest to analyze as it represents the usual and intended way a word or phrase is used. It is also more common in informational texts because it is used to state facts and definitions. While figurative language is typically associated with fiction and poetry, it can be found in informational texts as well. The reader must determine not only what is meant by the figurative language in context, but also how the author intended it to shape the overall text.

Inference in Informational Text

Inference refers to the reader's ability to understand the unwritten text, i.e., "read between the lines" in terms of an author's intent or message. The strategy asks that a reader not take everything he or she reads at face value but instead, add his or her own interpretation of what the author seems to be trying to convey. A reader's ability to make inferences relies on his or her ability to think clearly and logically about the text. It does not ask that the reader make wild speculation or guess about the material but demands that he or she be able to come to a sound conclusion about the material.

An author's use of less literal words and phrases requires readers to make more inference when they read. Since inference involves deduction—deriving conclusions from ideas assumed to be true—there's more room for interpretation. Still, critical readers who employ inference, if careful in their thinking, can still arrive at the logical, sound conclusions the author intends.

Textual Evidence in Informational Text

Once a reader has determined an author's thesis or main idea, he or she will need to understand how textual evidence supports interpretation of that thesis or main idea. Test takers will be asked direct questions regarding an author's main idea and may be asked to identify evidence that would support those ideas. This will require test takers to comprehend literal and figurative meanings within the text passage, be able to draw inferences from provided information, and be able to separate important evidence from minor supporting detail. It's often helpful to skim test questions and answer options prior to critically reading informational text; however, test takers should avoid the temptation to solely look for the correct answers. Just trying to find the "right answer" may cause test takers to miss important supporting textual evidence. Making mental note of test questions is only helpful as a guide when reading.

After identifying an author's thesis or main idea, a test taker should look at the supporting details that the author provides to back up his or her assertions, identifying those additional pieces of information that help expand the thesis. From there, test takers should examine the additional information and related details for credibility, the author's use of outside sources, and be able to point to direct evidence that supports the author's claims. It's also imperative that test takers be able to identify what is strong support and what is merely additional information that is nice to know but not necessary. Being able to make this differentiation will help test takers effectively answer questions regarding an author's use of supporting evidence within informational text.

Understanding Organizational Patterns and Structures

Organizational Structure within Informational Text

Informational text is specifically designed to relate factual information, and although it is open to a reader's interpretation and application of the facts, the structure of the presentation is carefully designed to lead the reader to a particular conclusion or central idea. When reading informational text, it is important that readers are able to understand its organizational structure as the structure often directly relates to an author's intent to inform and/or persuade the reader.

The first step in identifying the text's structure is to determine the thesis or main idea. The thesis statement and organization of a work are closely intertwined. A thesis statement indicates the writer's purpose and may include the scope and direction of the text. It may be presented at the beginning of a text or at the end, and it may be explicit or implicit.

Once a reader has a grasp of the thesis or main idea of the text, he or she can better determine its organizational structure. Test takers are advised to read informational text passages more than once in order to comprehend the material fully. It is also helpful to examine any text features present in the text including the table of contents, index, glossary, headings, footnotes, and visuals. The analysis of these features and the information presented within them, can offer additional clues about the central idea and structure of a text. The following questions should be asked when considering structure:

- How does the author assemble the parts to make an effective whole argument?
- Is the passage linear in nature and if so, what is the timeline or thread of logic?
- What is the presented order of events, facts, or arguments? Are these effective in contributing to the author's thesis?
- How can the passage be divided into sections? How are they related to each other and to the main idea or thesis?
- What key terms are used to indicate the organization?

Next, test takers should skim the passage, noting the first line or two of each body paragraph—the **topic sentences**—and the conclusion. Key **transitional terms**, such as on the other hand, also, because, however, therefore, most importantly, and first, within the text can also signal organizational structure. Based on these clues, readers should then be able to identify what type of organizational structure is being used. The following organizational structures are most common:

- *Problem/solution*—organized by an analysis/overview of a problem, followed by potential solution(s)
- *Cause/effect*—organized by the effects resulting from a cause or the cause(s) of a particular effect
- *Spatial order*—organized by points that suggest location or direction—e.g., top to bottom, right to left, outside to inside
- *Chronological/sequence order*—organized by points presented to indicate a passage of time or through purposeful steps/stages
- *Comparison/Contrast*—organized by points that indicate similarities and/or differences between two things or concepts
- *Order of importance*—organized by priority of points, often most significant to least significant or vice versa

Workplace and Community Documents

Workplace and community documents help employers to communicate within the business world and foster positive community relations outside of it. Workplace communications typically craft a specific message to a targeted audience while community documents send a broader or more generic message to a wider range of recipients.

Workplace Documents

Even though workplace-related documents are generated in a multitude of paper and electronic formats—memorandums, bulletin boards, presentations, web conferencing, instant messaging, and e-mails—in general all effective business communications share relevant information concisely, accurately, and purposefully. Supervisors rely on workplace documents to communicate expectations to subordinates (downward communication), and subordinates rely on workplace documents to submit progress reports, ask questions, and address concerns with their supervisors (upward communication).

- Memorandums: Designed to communicate information to a wide audience, memorandums inform staff of company-wide policy changes. Similar to an e-mail, a memorandum has a header near the top which identifies the intended audience, the author of the memo, the subject of the memo, and the date it was issued. Unlike an e-mail, though, memorandums are longer, can be submitted in paper or electronic form, and contain an introduction that identifies the topic or problem, a body that expands on the topic, and a conclusion that suggests a course of action or solution.

- Bulletin Boards: Regardless of whether they are in paper or electronic form, bulletin boards provide a less formal setting for supervisors and staff to communicate. Bulletin boards are a perfect medium to post federal and state regulations, employee incentive initiatives, volunteer opportunities, and company news. While paper bulletin boards are limited to a specific office and personnel, electronic bulletin boards have the capability of broadcasting information nationally and even globally.

- Presentations: Presentations can be created with a variety of software—*PowerPoints*, *Google Slides*, and *Prezi*—and are given extemporaneously. Typically, presentations have an introductory slide, informational slides, and a concluding slide that gives the presenter's audience the opportunity to ask questions or create a dialogue. Presentations relay information in a media-rich format: graphics, tables, and hyperlinked documents and videos are easily imbedded within the slides.

- Web Conferencing: Web conferencing allows for employees to collaborate on projects and tasks. Employees are able to talk or videoconference from different locations, making it possible for remote workers from around the globe to participate simultaneously in one meeting. Web conferencing can be done via telephone with visuals (*PowerPoint* or *Microsoft Word* documents) or via video camera programs like *WebEx®*, *PGI GlobalMeet®*, and *Skype®*.

- Instant Messaging: Many staff outside of centralized locations who cannot communicate verbally with coworkers or supervisors rely, instead, on instant messaging programs. Instant messaging programs can deliver messages one-on-one or in circles and groups, and some software even provides screen-sharing capabilities. Since instant messaging is faster than e-mail, for many employees it has become the preferred method of communicating over long distances.

- E-Mails: In today's fast-paced business world, e-mails are heavily relied upon because they provide a platform that is perfect for quickly communicating brief, concise messages to targeted audiences. E-mails not only detail who the sender and receiver are but also provide a date, time, and subject line. Unlike a memorandum, sender and receiver can communicate back and forth, and, more importantly, they may do so over long distances. Many businesses today rely on *Google*, *Yahoo*, and *Outlook* mail severs.

<u>Community Documents</u>

Where business documents target a specific audience and often contain a higher level of proprietary, confidential, or sensitive information, community documents invite larger groups to discuss business matters in a less restrictive environment. Because community documents often help bridge the gap between businesses and community, an effective newsletter, discussion board, blog, website, and app have the power to influence public perception of private companies.

- Newsletters: Newsletters are used to provide the public information about the business. It can be used to generate excitement, inform or persuade staff or consumers, or give tips on how the public can contact or work with businesses. Newsletters can be mailed or sent electronically. Generally, newsletters are sent weekly, monthly, or quarterly. They can be interactive, providing the community a glance at what is going on with a business that they interact with, and, moreover, newsletters can communicate a company's mission, values, and priorities.

- Discussion Boards: Discussion boards offer a place to go to share information on a specific topic. Discussion boards are organized on menus, submenus, and discussion threads. People visit discussion boards to find out more about a topic. Discussion board members are granted greater access to the site and have greater power to publish and comment, but, still, visitors are welcome, and they often get difficult or obscure questions resolved.

- Blogs: A blog, which is usually centered on a specific topic or theme, is a website where individuals post and update information constantly. Blogs tend to feature the newest posts first while archiving older ones. Articles, editorials, images, videos, surveys, and social media (just to name a few) can all be imbedded within blogs. Menus, sidebars, recent posts, and search boxes help visitors wade through a dizzying array of media formats and topics. Though individuals can hire a web designer to create a blog for them, most people use existing platforms like *WordPress®*, *Blogger®*, and *Tumbler®*.

- Websites: As the world becomes more technologically savvy, a website can be used to house community documents, giving consumers instant access to tools they will need to interact with the business. It can house forms, contact information, discussion boards, surveys, and blogs. It is a one-stop-shop that can assist in the interaction between the business and the consumer. Websites are advantageous because they can be accessed via computer, tablet, or mobile phone.

- Apps: Similar to a website, an application (app) gives businesses yet another method to reach individuals or segments of a community. Apps provide instant access to forms and other documentation. Designed to be downloaded on tablets or mobile phones, apps are streamlined and intuitive, allowing consumers on the go to access information at their convenience. The primary operating systems for apps include *iOS®* and *Android®*.

Identifying Primary Sources in Media

Primary sources are best defined as records or items that serve as historical evidence. To be considered primary, the source documents or objects must have been created during the time period in which they reference. Examples include diaries, newspaper articles, speeches, government documents, photographs, and historical artifacts. In today's digital age, primary sources, which were once in print, often are embedded in secondary sources. Secondary sources, such as websites, history books, databases, or reviews, contain analysis or commentary on primary sources. **Secondary sources** borrow information from primary sources through the process of quoting, summarizing, or paraphrasing.

Today's students often complete research online through electronic sources. Electronic sources offer advantages over print and can be accessed on virtually any computer, while libraries or other research centers are limited to fixed locations and specific catalogs. Electronic sources are also efficient and yield massive amounts of data in seconds. The user can tailor a search based on key words, publication years, and article length. Lastly, many databases provide the user with instant citations, saving the user the trouble of manually assembling sources.

Though electronic sources yield powerful results, researchers must use caution. While there are many reputable and reliable sources on the internet, just as many are unreliable or biased sources. It's up to the researcher to examine and verify the reliability of sources. *Wikipedia*, for example, may or may not be accurate, depending on the contributor. Many databases, such as *EBSCO* or *SIRS*, offer peer reviewed articles, meaning the publications have been reviewed for the quality of their content.

Evaluating an Argument and its Claims

It's important to evaluate the author's supporting details to be sure that the details are credible, provide evidence of the author's point, and directly support the main idea. Though shocking statistics grab readers' attention, their use could be ineffective information in the piece. Details like this are crucial to understanding the passage and evaluating how well the author presents their argument and evidence.

Readers draw **conclusions** about what an author has presented. This helps them better understand what the writer has intended to communicate and whether or not they agree with what the author has offered. There are a few ways to determine a logical conclusion, but careful reading is the most important. It's helpful to read a passage a few times, noting details that seem important to the piece. Sometimes, readers arrive at a conclusion that is different than what the writer intended or come up with more than one conclusion.

Evidence

It is important to distinguish between *fact and opinion* when reading a piece of writing. When an author presents **facts**, such as statistics or data, readers should be able to check those facts and make sure they are accurate. When authors use **opinion**, they are sharing their own thoughts and feelings about a subject.

Textual evidence helps readers draw a conclusion about a passage. Textual evidence refers to information—facts and examples that support the main point; it will likely come from outside sources and can be in the form of quoted or paraphrased material. In order to draw a conclusion from evidence, it's important to examine the credibility and validity of that evidence as well as how (and if) it relates to the main idea.

Credibility

Critical readers examine the facts used to support an author's argument. They check the facts against other sources to be sure those facts are correct. They also check the validity of the sources used to be sure those sources are credible, academic, and/or peer-reviewed. Consider that when an author uses another person's opinion to support their argument, even if it is an expert's opinion, it is still only an opinion and should not be taken as fact. A strong argument uses valid, measurable facts to support ideas. Even then, the reader may disagree with the argument as it may be rooted in their personal beliefs.

An authoritative argument may use the facts to sway the reader. Because of this, a writer may choose to only use the information and expert opinion that supports their viewpoint.

Appeal to Emotion

An author's argument might also appeal to readers' emotions, perhaps by including personal stories and anecdotes (a short narrative of a specific event).

The next example presents an appeal to emotion. By sharing the personal anecdote of one student and speaking about emotional topics like family relationships, the author invokes the reader's empathy in asking them to reconsider the school rule.

> Our school should abolish its current ban on campus cell phone use. If students aren't able to use their phones during the school day, many of them feel isolated from their loved ones. For example, last semester, one student's grandmother had a heart attack in the morning. However, because he couldn't use his cell phone, the student didn't know about his grandmother's accident until the end of the day—when she had already passed away, and it was too late to say goodbye. By preventing students from contacting their friends and family, our school is placing undue stress and anxiety on students.

Counter-Arguments

If an author presents a differing opinion or a **counter-argument** in order to refute it, the reader should consider how and why the information is being presented. It is meant to strengthen the original argument and shouldn't be confused with the author's intended conclusion, but it should also be considered in the reader's final evaluation. On the contrary, sometimes authors will concede to an opposing argument by recognizing the validity the other side has to offer. A concession will allow readers to see both sides of the argument in an unbiased light, thereby increasing the credibility of the author.

Authors can also reflect **bias** if they ignore an opposing viewpoint or present their side in an unbalanced way. A strong argument considers the opposition and finds a way to refute it. Critical readers should look for an unfair or one-sided presentation of the argument and be skeptical, as a bias may be present. Even if this bias is unintentional, if it exists in the writing, the reader should be wary of the validity of the argument.

Revising/Editing

Parts of Speech

Nouns

A noun is a person, place, thing, or idea. All nouns fit into one of two types, common or proper.

A common noun is a word that identifies any of a class of people, places, or things. Examples include numbers, objects, animals, feelings, concepts, qualities, and actions. *A, an,* or *the* usually precedes the common noun. These parts of speech are called articles. Here are some examples of sentences using nouns preceded by articles.

A building is under construction.
The girl would like to move to *the* city.

A proper noun (also called a proper name) is used for the specific name of an individual person, place, or organization. The first letter in a proper noun is capitalized. "My name is *Mary*." "I work for *Walmart*."

Nouns sometimes serve as adjectives (which themselves describe nouns), such as "hockey player" and "state government."

Pronouns

A word used in place of a noun is known as a pronoun. Pronouns are words like *I, mine, hers,* and *us*.

Pronouns can be split into different classifications (seen below) which make them easier to learn; however, it's not important to memorize the classifications.

- Personal pronouns: refer to people
- First person: we, I, our, mine
- Second person: you, yours
- Third person: he, them
- Possessive pronouns: demonstrate ownership (mine, his, hers, its, ours, theirs, yours)
- Interrogative pronouns: ask questions (what, which, who, whom, whose)
- Relative pronouns: include the five interrogative pronouns and others that are relative (whoever, whomever, that, when, where)
- Demonstrative pronouns: replace something specific (this, that, those, these)
- Reciprocal pronouns: indicate something was done or given in return (each other, one another)
- Indefinite pronouns: have a nonspecific status (anybody, whoever, someone, everybody, somebody)

Indefinite pronouns such as *anybody, whoever, someone, everybody*, and *somebody* command a singular verb form, but others such as *all, none,* and *some* could require a singular or plural verb form.

Antecedents

An antecedent is the noun to which a pronoun refers; it needs to be written or spoken before the pronoun is used. For many pronouns, antecedents are imperative for clarity. In particular, a lot of the personal, possessive, and demonstrative pronouns need antecedents. Otherwise, it would be unclear who or what someone is referring to when they use a pronoun like *he* or *this*.

Pronoun reference means that the pronoun should refer clearly to one, clear, unmistakable noun (the antecedent).

Pronoun-antecedent agreement refers to the need for the antecedent and the corresponding pronoun to agree in gender, person, and number. Here are some examples:

> The *kidneys* (plural antecedent) are part of the urinary system. *They* (plural pronoun) serve several roles.

> The kidneys are part of the *urinary system* (singular antecedent). *It* (singular pronoun) is also known as the renal system.

Pronoun Cases

The subjective pronouns —*I, you, he/she/it, we, they,* and *who*—are the subjects of the sentence.

> Example: *They* have a new house.

The objective pronouns—*me, you* (*singular*), *him/her, us, them,* and *whom*—are used when something is being done for or given to someone; they are objects of the action.

> Example: The teacher has an apple for *us*.

The possessive pronouns—*mine, my, your, yours, his, hers, its, their, theirs, our,* and *ours*—are used to denote that something (or someone) belongs to someone (or something).

> Example: It's *their* chocolate cake.
> Even Better Example: It's *my* chocolate cake!

One of the greatest challenges and worst abuses of pronouns concerns *who* and *whom*. Just knowing the following rule can eliminate confusion. *Who* is a subjective-case pronoun used only as a subject or subject complement. *Whom* is only objective-case and, therefore, the object of the verb or preposition.

> *Who* is going to the concert?

> You are going to the concert with *whom*?

Hint: When using *who* or *whom*, think of whether someone would say *he* or *him*. If the answer is *he*, use *who*. If the answer is *him*, use *whom*. This trick is easy to remember because *he* and *who* both end in vowels, and *him* and *whom* both end in the letter *M*.

Adjectives

"The *extraordinary* brain is the *main* organ of the central nervous system." The adjective *extraordinary* describes the brain in a way that causes one to realize it is more exceptional than some of the other organs while the adjective *main* defines the brain's importance in its system.

An *adjective* is a word or phrase that names an attribute that describes or clarifies a noun or pronoun. This helps the reader visualize and understand the characteristics—size, shape, age, color, origin, etc.—of a person, place, or thing that otherwise might not be known. Adjectives breathe life, color, and depth into the subjects they define.

Adjectives often precede the nouns they describe.

> *She drove her <u>new</u> car.*

However, adjectives can also come later in the sentence.

> *Her car is <u>new</u>.*

Adjectives using the prefix *a–* can only be used after a verb.

> Correct: The dog was alive until the car ran up on the curb and hit him.
> Incorrect: The alive dog was hit by a car that ran up on the curb.

Other examples of this rule include *awake, ablaze, ajar, alike,* and *asleep*.

Other adjectives used after verbs concern states of health.

> The girl was finally *well* after a long bout of pneumonia.
> The boy was *fine* after the accident.

An adjective phrase is not a bunch of adjectives strung together, but a group of words that describes a noun or pronoun and, thus, functions as an adjective. Very happy is an adjective phrase; so are way too hungry and passionate about traveling.

Possessives

In grammar, possessive nouns show ownership, which was seen in previous examples like *mine, yours,* and *theirs*.

Singular nouns are generally made possessive with an apostrophe and an *s* (*'s*).

> My *uncle's* new car is silver.
> The *dog's* bowl is empty.
> *James's* ties are becoming outdated.

Plural nouns ending in *s* are generally made possessive by just adding an apostrophe (*'*):

> The pistachio nuts' saltiness is added during roasting. (The saltiness of pistachio nuts is added during roasting.)
>
> The students' achievement tests are difficult. (The achievement tests of the students are difficult.)

If the plural noun does not end in an *s* such as *women,* then it is made possessive by adding an apostrophe *s* (*'s*)—*women's.*

Indefinite possessive pronouns such as *nobody* or *someone* become possessive by adding an apostrophe *s*— *nobody's* or *someone's.*

Verbs

A verb is the part of speech that describes an action, state of being, or occurrence.

A verb forms the main part of a predicate of a sentence. This means that the verb explains what the noun (which will be discussed shortly) is doing. A simple example is *time <u>flies</u>*. The verb *flies* explains what the action of the noun, *time*, is doing. This example is a *main* verb.

Helping (auxiliary) verbs are words like *have, do, be, can, may, should, must,* and *will.* "I *should* go to the store." Helping verbs assist main verbs in expressing tense, ability, possibility, permission, or obligation.

Particles are minor function words like *not, in, out, up,* or *down* that become part of the verb itself. "I might *not*."

Participles are words formed from verbs that are often used to modify a noun, noun phrase, verb, or verb phrase.

> The *running* teenager collided with the cyclist.

Participles can also create compound verb forms.

> He is *speaking*.

Verbs have five basic forms: the *base* form, the *-s* form, the *-ing* form, the *past* form, and the *past participle* form.

The past forms are either regular (*love/loved; hate/hated*) or irregular because they don't end by adding the common past tense suffix "-ed" (*go/went; fall/fell; set/set*).

Adverbs

Adverbs have more functions than adjectives because they modify or qualify verbs, adjectives, or other adverbs as well as word groups that express a relation of place, time, circumstance, or cause. Therefore, adverbs answer any of the following questions: *How, when, where, why, in what way, how often, how much, in what condition,* and/or *to what degree. How good looking is he? He is very handsome.*

Here are some examples of adverbs for different situations:

- how: quickly
- when: daily
- where: there
- in what way: easily
- how often: often
- how much: much
- in what condition: badly
- what degree: hardly

As one can see, for some reason, many adverbs end in *-ly.*

Adverbs do things like emphasize (*really, simply,* and *so*), amplify (*heartily, completely,* and *positively*), and tone down (*almost, somewhat,* and *mildly*).

Adverbs also come in phrases.

> *The dog ran as though his life depended on it.*

Prepositions

Prepositions are connecting words and, while there are only about 150 of them, they are used more often than any other individual groups of words. They describe relationships between other words. They are placed before a noun or pronoun, forming a phrase that modifies another word in the sentence.

Prepositional phrases begin with a preposition and end with a noun or pronoun, the object of the preposition. *A pristine lake is near the store and behind the bank.*

Some commonly used prepositions are *about, after, anti, around, as, at, behind, beside, by, for, from, in, into, of, off, on, to,* and *with.*

Complex prepositions, which also come before a noun or pronoun, consist of two or three words such as *according to, in regards to,* and *because of.*

Interjections

Interjections are words used to express emotion. Examples include *wow*, *ouch*, and *hooray*. Interjections are often separate from sentences; in those cases, the interjection is directly followed by an exclamation point. In other cases, the interjection is included in a sentence and followed by a comma. The punctuation plays a big role in the intensity of the emotion that the interjection is expressing. Using a comma or semicolon indicates less excitement than using an exclamation mark.

Conjunctions

Conjunctions are vital words that connect words, phrases, thoughts, and ideas. Conjunctions show relationships between components. There are two types:

Coordinating conjunctions are the primary class of conjunctions placed between words, phrases, clauses, and sentences that are of equal grammatical rank; the coordinating conjunctions are *for*, *and*, *nor*, *but*, *or*, *yet*, and *so*. A useful memorization trick is to remember that the first letter of these conjunctions collectively spell the word *fanboys*.

> I need to go shopping, *but* I must be careful to leave enough money in the bank.
> She wore a black, red, *and* white shirt.

Subordinating conjunctions are the secondary class of conjunctions. They connect two unequal parts, one main (or independent) and the other subordinate (or dependent). I must go to the store *even though* I do not have enough money in the bank.

> *Because* I read the review, I do not want to go to the movie.

Notice that the presence of subordinating conjunctions makes clauses dependent. *I read the review* is an independent clause, but *because* makes the clause dependent. Thus, it needs an independent clause to complete the sentence.

Punctuation

Periods (.) are used to end a sentence that is a statement (declarative) or a command (imperative). They should not be used in a sentence that asks a question or is an exclamation. Periods are also used in abbreviations, which are shortened versions of words.

- Declarative: The boys refused to go to sleep.
- Imperative: Walk down to the bus stop.
- Abbreviations: Joan Roberts, M.D., Apple Inc., Mrs. Adamson
- If a sentence ends with an abbreviation, it is inappropriate to use two periods. It should end with a single period after the abbreviation.

  The chef gathered the ingredients for the pie, which included apples, flour, sugar, etc.

Question marks *(?)* are used with direct questions (interrogative). An indirect question can use a period:

> Interrogative: When does the next bus arrive?
>
> Indirect Question: I wonder when the next bus arrives.

An exclamation point *(!)* is used to show strong emotion or can be used as an interjection. This punctuation should be used sparingly in formal writing situations.

> What an amazing shot!
>
> Whoa!

In a sentence, colons are used before a list, a summary or elaboration, or an explanation related to the preceding information in the sentence:

> There are two ways to reserve tickets for the performance: by phone or in person.
>
> One thing is clear: students are spending more on tuition than ever before.

As these examples show, a colon must be preceded by an independent clause. However, the information after the colon may be in the form of an independent clause or in the form of a list.

Semicolons can be used in two different ways—to join ideas or to separate them. In some cases, semicolons can be used to connect what would otherwise be stand-alone sentences. Each part of the sentence joined by a semicolon must be an independent clause. The use of a semicolon indicates that these two independent clauses are closely related to each other:

> The rising cost of childcare is one major stressor for parents; healthcare expenses are another source of anxiety.
>
> Classes have been canceled due to the snowstorm; check the school website for updates.

Semicolons can also be used to divide elements of a sentence in a more distinct way than simply using a comma. This usage is particularly useful when the items in a list are especially long and complex and contain other internal punctuation.

> Retirees have many modes of income: some survive solely off their retirement checks; others supplement their income through part time jobs, like working in a supermarket or substitute teaching; and others are financially dependent on the support of family members, friends, and spouses.

Subject-Verb Agreement

The subject of a sentence and its verb must agree. The cornerstone rule of subject-verb agreement is that subject and verb must agree in number. Whether the subject is singular or plural, the verb must follow suit.

> Incorrect: The houses is new.
> Correct: The houses are new.
> Also Correct: The house is new.

In other words, a singular subject requires a singular verb; a plural subject requires a plural verb. The words or phrases that come between the subject and verb do not alter this rule.

> Incorrect: The houses built of brick is new.
> Correct: The houses built of brick are new.
>
> Incorrect: The houses with the sturdy porches is new.
> Correct: The houses with the sturdy porches are new.

The subject will always follow the verb when a sentence begins with *here* or *there.* Identify these with care.

> Incorrect: Here *is* the *houses* with sturdy porches.
> Correct: Here *are* the *houses* with sturdy porches.

The subject in the sentences above is not *here*, it is *houses*. Remember, *here* and *there* are never subjects. Be careful that contractions such as *here's* or *there're* do not cause confusion!

Two subjects joined by *and* require a plural verb form, except when the two combine to make one thing:

> Incorrect: Garrett and Jonathan is over there.
> Correct: Garrett and Jonathan are over there.
>
> Incorrect: Spaghetti and meatballs are a delicious meal!
> Correct: Spaghetti and meatballs is a delicious meal!

In the example above, *spaghetti and meatballs* is a compound noun. However, *Garrett and Jonathan* is not a compound noun.

Two singular subjects joined by *or, either/or,* or *neither/nor* call for a singular verb form.

Incorrect: Butter or syrup are acceptable.
Correct: Butter or syrup is acceptable.

Plural subjects joined by *or*, *either/or*, or *neither/nor* are, indeed, plural.

The chairs or the boxes are being moved next.

If one subject is singular and the other is plural, the verb should agree with the closest noun.

Correct: The chair or the boxes are being moved next.
Correct: The chairs or the box is being moved next.

Some plurals of money, distance, and time call for a singular verb.

Incorrect: Three dollars *are* enough to buy that.
Correct: Three dollars *is* enough to buy that.

For words declaring degrees of quantity such as *many of, some of,* or *most of,* let the noun that follows *of* be the guide:

Incorrect: Many of the books is in the shelf.
Correct: Many of the books are in the shelf.

Incorrect: Most of the pie *are* on the table.
Correct: Most of the pie *is* on the table.

For indefinite pronouns like anybody or everybody, use singular verbs.

Everybody *is* going to the store.

However, the pronouns *few, many, several, all, some,* and *both* have their own rules and use plural forms.

Some *are* ready.

Some nouns like *crowd* and *congress* are called collective nouns, and they require a singular verb form.

Congress *is* in session.
The news *is* over.

Books and movie titles, though, including plural nouns such as *Great Expectations*, also require a singular verb. Remember that only the subject affects the verb. While writing tricky subject-verb arrangements, say them aloud. Listen to them. Once the rules have been learned, one's ear will become sensitive to them, making it easier to pick out what's right and what's wrong.

Independent and Dependent Clauses

Independent and dependent clauses are strings of words that contain both a subject and a verb. An independent clause can stand alone as complete thought, but a dependent clause cannot. A dependent clause relies on other words to be a complete sentence.

> Independent clause: The keys are on the counter.
> Dependent clause: If the keys are on the counter

Notice that both clauses have a subject (*keys*) and a verb (*are*). The independent clause expresses a complete thought, but the word *if* at the beginning of the dependent clause makes it dependent on other words to be a complete thought.

> Independent clause: If the keys are on the counter, please give them to me.

This example constitutes a complete sentence since it includes at least one verb and one subject and is a complete thought. In this case, the independent clause has two subjects (*keys* & an implied *you*) and two verbs (*are* & *give*).

> Independent clause: I went to the store.
> Dependent clause: Because we are out of milk,
>
> Complete Sentence: Because we are out of milk, I went to the store.
> Complete Sentence: I went to the store because we are out of milk.

A phrase is a group of words that does not contain both a subject and a verb.

Word Usage

Word usage*:* the way and manner in which writers choose to use words (or phrases). This is a vital consideration in order to create excellence in writing. Context plays a role in the selection of words, as does simple choice.

Correct word usage can be as basic and imperative as making the right grammatical choice.

> Incorrect: I never *seen* that play
> Correct: I never *saw* that play.

Seen (past participle of see) in place of *saw* (the correct past tense of *see*) is a relatively common and grievous mistake.

> Incorrect: *There* going to the store with *there* parents.
> Correct: *They're* going to the store with *their* parents.

There (adverb) mistaken for *they're* (contraction for "they are") and *there* (adverb) for *their* (pronoun) are also frequent grammatical errors.

Correct word usage can be as simple as choosing between synonyms. For instance, determining whether to say *for instance* or *for example*! Would it be better to use the word *incongruous* or *inappropriate* for *unsuited,* or when the woman is *uncouth,* would *redneck, uninformed,* or *illiterate* better define her?

Correct word usage can also be as simple as knowing how to access and make good use of dictionaries, thesauruses, grammar correction software, and search engines on computers, tablets, and smartphones.

Context Clues

Context clues help readers understand unfamiliar words, and thankfully, there are many types.

Synonyms are words or phrases that have nearly, if not exactly, the same meaning as other words or phrases

Large boxes are needed to pack *big* items.

Antonyms are words or phrases that have opposite definitions. Antonyms, like synonyms, can serve as context clues, although more cryptically.

Large boxes are not needed to pack *small* items.

Definitions are sometimes included within a sentence to define uncommon words.

They practiced the *rumba*, a *type of dance*, for hours on end.

Explanations provide context through elaboration.

Large boxes holding items weighing over 60 pounds were stacked in the corner.

Contrast provides ways in which things are different.

These *minute* creatures were much different than the *huge* mammals that the zoologist was accustomed to dealing with.

Word Parts

By analyzing and understanding Latin, Greek, and Anglo-Saxon word roots, prefixes, and suffixes one can better understand word meanings. Of course, people can always look words up if a dictionary or thesaurus, if available, but meaning can often be gleaned on the spot if the writer learns to dissect and examine words.

A word can consist of the following:

- root
- root + suffix
- prefix + root
- prefix + root + suffix

For example, if someone was unfamiliar with the word *submarine* they could break the word into its parts.

prefix + root
sub + marine

It can be determined that *sub* means *below* as in *subway* and *subpar*. Additionally, one can determine that *marine* refers to *the sea* as in *marine life*. Thus, it can be figured that *submarine* refers to something below the water.

Roots

Roots are the basic components of words. Many roots can stand alone as individual words, but others must be combined with a prefix or suffix to be a word. For example, *calc* is a root but it needs a suffix to be an actual word (*calcium*).

Prefixes

A *prefix* is a word, letter, or number that is placed before another. It adjusts or qualifies the root word's meaning. When written alone, prefixes are followed by a dash to indicate that the root word follows. Some of the most common prefixes are the following:

Prefix	Meaning	Example
dis-	not or opposite of	disabled
in-, im-, il-, ir-	not	illiterate
re-	again	return
un-	not	unpredictable
anti-	against	antibacterial
fore-	before	forefront
mis-	wrongly	misunderstand
non-	not	nonsense
over-	more than normal	overabundance
pre-	before	preheat
super-	above	superman

Suffixes

A suffix is a letter or group of letters added at the end of a word to form another word. The word created from the root and suffix is either a different tense of the same root (*help + ed = helped*) or a new word (*help + ful = helpful*). When written alone, suffixes are preceded by a dash to indicate that the root word comes before.

Some of the most common suffixes are the following:

Suffix	Meaning	Example
ed	makes a verb past tense	wash*ed*
ing	makes a verb a present participle verb	wash*ing*
ly	to make characteristic of	love*ly*
s/es	to make more than one	chairs, box*es*
able	can be done	deplor*able*
al	having characteristics of	comic*al*
est	comparative	great*est*
ful	full of	wonder*ful*
ism	belief in	commun*ism*
less	without	faithless
ment	action or process	accomplish*ment*
ness	state of	happi*ness*
ize, ise	to render, to make	steril*ize*, advert*ise*
cede/ceed/sede	go	concede, proceed, supersede

Here are some helpful tips:

- When adding a suffix that starts with a vowel (for example, *-ed*) to a one-syllable root whose vowel has a short sound and ends in a consonant (for example, *stun*), double the final consonant of the root (*n*).

  stun + ed = stun*n*ed

  Exception: If the past tense verb ends in *x* such as *box*, do not double the *x*.

  box + ed = boxed

- If adding a suffix that starts with a vowel (*-er*) to a multi-syllable word ending in a consonant (*begin*), double the consonant (*n*).

  begin + er = begin*n*er

- If a short vowel is followed by two or more consonants in a word such as *i+t+c+h = itch,* do <u>not</u> double the last consonant.

  itch + ed = itched

- If adding a suffix that starts with a vowel (*-ing*) to a word ending in *e* (for example*, name*), that word's final *e* is generally (but not always) dropped.

  name + ing = naming
  exception: manage + able = manag*e*able

- If adding a suffix that starts with a consonant (*-ness*) to a word ending in *e* (*complete*), the *e* generally (but not always) remains.

  complete + ness = completeness
  exception: judge + ment = judgment

- There is great diversity on handling words that end in y. For words ending in a vowel + y, nothing changes in the original word.

  play + ed = played

- For words ending in a consonant + *y*, change the *y* to *i* when adding any suffix except for *–ing*.

  marry + ed = married
  marry + ing = marrying

Sentence Fluency

It's time to take what's been studied and put it all together in order to construct well-written sentences and paragraphs that have correct structure. Learning and utilizing the mechanics of structure will encourage effective, professional results, and adding some creativity will elevate one's writing to a higher level.

First, let's review the basic elements of sentences.

A sentence is a set of words that make up a grammatical unit. The words must have certain elements and be spoken or written in a specific order to constitute a complete sentence that makes sense.

1. A sentence must have a subject (a noun or noun phrase). The subject tells whom or what the sentence is addressing (i.e. what it is about).

2. A sentence must have an action or state of being (*a verb*). To reiterate: A verb forms the main part of the predicate of a sentence. This means that it explains what the noun is doing.

3. A sentence must convey a complete thought.

When examining writing, be mindful of grammar, structure, spelling, and patterns. Sentences can come in varying sizes and shapes; so, the point of grammatical correctness is not to stamp out creativity or diversity in writing. Rather, grammatical correctness ensures that writing will be enjoyable and clear. One of the most common methods for catching errors is to mouth the words as you read them. Many typos are fixed automatically by our brain, but mouthing the words often circumvents this instinct and helps one read what's actually on the page. Often, grammar errors are caught not by memorization of grammar rules but by the training of one's mind to know whether something sounds right or not.

A sentence fragment is a failed attempt to create a complete sentence because it's missing a required noun or verb. Fragments don't function properly because there isn't enough information to understand the writer's intended meaning. For example:

> Seat belt use corresponds to a lower rate of hospital visits, reducing strain on an already overburdened healthcare system. Insurance claims as well.

Look at the last sentence: *Insurance claims as well*. What does this mean? This is a fragment because it has a noun but no verb, and it leaves the reader guessing what the writer means about insurance claims. Many readers can probably infer what the writer means, but this distracts them from the flow of the writer's argument. Choosing a suitable replacement for a sentence fragment may be one of the questions on the test. The fragment is probably related to the surrounding content, so look at the overall point the writer is trying to make and choose the answer that best fits that idea.

Remember that sometimes a fragment can look like a complete sentence or have all the nouns and verbs it needs to make sense. Consider the following two examples:

> Seat belt use corresponds to a lower rate of hospital visits.
>
> Although seat belt use corresponds to a lower rate of hospital visits.

Both examples above have nouns and verbs, but only the first sentence is correct. The second sentence is a fragment, even though it's actually longer. The key is the writer's use of the word *although*. Starting a sentence with although turns that part into a subordinate clause (more on that next). Keep in mind that one doesn't have to remember that it's called a subordinate clause on the test. Just be able to recognize that the words form an incomplete thought and identify the problem as a sentence fragment.

A run-on sentence is, in some ways, the opposite of a fragment. It contains two or more sentences that have been improperly forced together into one. An example of a run-on sentence looks something like this:

> Seat belt use corresponds to a lower rate of hospital visits it also leads to fewer insurance claims.

Here, there are two separate ideas in one sentence. It's difficult for the reader to follow the writer's thinking because there is no transition from one idea to the next. On the test, choose the best way to correct the run-on sentence.

Here are two possibilities for the sentence above:

> Seat belt use corresponds to a lower rate of hospital visits. It also leads to fewer insurance claims.
>
> Seat belt use corresponds to a lower rate of hospital visits, but it also leads to fewer insurance claims.

Both solutions are grammatically correct, so which one is the best choice? That depends on the point that the writer is trying to make. Always read the surrounding text to determine what the writer wants to demonstrate, and choose the option that best supports that thought.

Another type of run-on occurs when writers use inappropriate punctuation:

> This winter has been very cold, some farmers have suffered damage to their crops.

Though a comma has been added, this sentence is still not correct. When a comma alone is used to join two independent clauses, it is known as a comma splice. Without an appropriate conjunction, a comma cannot join two independent clauses by itself.

Types of Sentences

There isn't an overabundance of absolutes in grammar, but here is one: every sentence in the English language falls into one of four categories.

- Declarative: a simple statement that ends with a period

    The price of milk per gallon is the same as the price of gasoline.

- Imperative: a command, instruction, or request that ends with a period

    Buy milk when you stop to fill up your car with gas.

- Interrogative: a question that ends with a question mark

    Will you buy the milk?

- Exclamatory: a statement or command that expresses emotions like anger, urgency, or surprise and ends with an exclamation mark

    Buy the milk now!

Declarative sentences are the most common type, probably because they are comprised of the most general content, without any of the bells and whistles that the other three types contain. They are, simply, declarations or statements of any degree of seriousness, importance, or information.

Imperative sentences often seem to be missing a subject. The subject is there, though; it is just not visible or audible because it is implied. Look at the imperative example sentence.

Buy the milk when you fill up your car with gas.

You is the implied subject, the one to whom the command is issued. This is sometimes called *the understood you* because it is understood that *you* is the subject of the sentence.

Interrogative sentences—those that ask questions—are defined as such from the idea of the word interrogation, the action of questions being asked of suspects by investigators. Although that is serious business, interrogative sentences apply to all kinds of questions.

To exclaim is at the root of exclamatory sentences. These are made with strong emotions behind them. The only technical difference between a declarative or imperative sentence and an exclamatory one is the exclamation mark at the end. The example declarative and imperative sentences can both become an exclamatory one simply by putting an exclamation mark at the end of the sentences.

The price of milk per gallon is the same as the price of gasoline!
Buy milk when you stop to fill up your car with gas!

After all, someone might be really excited by the price of gas or milk, or they could be mad at the person that will be buying the milk! However, as stated before, exclamation marks in abundance defeat their own purpose! After a while, they begin to cause fatigue! When used only for their intended purpose, they can have their expected and desired effect.

Parallel Structure in a Sentence

Parallel structure, also known as parallelism, refers to using the same grammatical form within a sentence. This is important in lists and for other components of sentences.

> Incorrect: At the recital, the boys and girls were dancing, singing, and played musical instruments.
> Correct: At the recital, the boys and girls were dancing, singing, and playing musical instruments.

Notice that in the second example, *played* is not in the same verb tense as the other verbs, nor is it compatible with the helping verb *were*. To test for parallel structure in lists, try reading each item as if it were the only item in the list.

> The boys and girls were dancing.
> The boys and girls were singing.
> The boys and girls were played musical instruments.

Suddenly, the error in the sentence becomes very clear. Here's another example:

> Incorrect: After the accident, I informed the police *that Mrs. Holmes backed* into my car, *that Mrs. Holmes got out* of her car to look at the damage, and *she was driving* off without leaving a note.
>
> Correct: After the accident, I informed the police *that Mrs. Holmes backed* into my car, *that Mrs. Holmes got out* of her car to look at the damage, and *that Mrs. Holmes drove off* without leaving a note.
>
> Correct: After the accident, I informed the police that Mrs. Holmes *backed* into my car, *got out* of her car to look at the damage, and *drove off* without leaving a note.

Note that there are two ways to fix the nonparallel structure of the first sentence. The key to parallelism is consistent structure.

Sentence Structures

A simple sentence has one independent clause.

> I am going to win.

A compound sentence has two independent clauses. A conjunction—*for, and, nor, but, or, yet, so*—links them together. Note that each of the independent clauses has a subject and a verb.

> I am going to win, but the odds are against me.

A complex sentence has one independent clause and one or more dependent clauses.

> I am going to win, even though I don't deserve it.

Even though I don't deserve it is a dependent clause. It does not stand on its own. Some conjunctions that link an independent and a dependent clause are *although, because, before, after, that, when, which*, and *while*.

A compound-complex sentence has at least three clauses, two of which are independent and at least one that is a dependent clause.

While trying to dance, I tripped over my partner's feet, but I regained my balance quickly.

The dependent clause is *While trying to dance*.

Forming Paragraphs

A good paragraph should have the following characteristics:

- Be logical with organized sentences
- Have a unified purpose within itself
- Use sentences as building blocks
- Be a distinct section of a piece of writing
- Present a single theme introduced by a topic sentence
- Maintain a consistent flow through subsequent, relevant, well-placed sentences
- Tell a story of its own or have its own purpose, yet connect with what is written before and after
- Enlighten, entertain, and/or inform

Though certainly not set in stone, the length should be a consideration for the reader's sake, not merely for the sake of the topic. When paragraphs are especially short, the reader might experience an irregular, uneven effect; when they're much longer than 250 words, the reader's attention span, and probably their retention, is challenged. While a paragraph can technically be a sentence long, a good rule of thumb is for paragraphs to be at least three sentences long and no more than ten sentence long. An optimal word length is 100 to 250 words.

Coherent Paragraphs

Coherence is simply defined as the quality of being logical and consistent. In order to have coherent paragraphs, therefore, authors must be logical and consistent in their writing, whatever the document might be. Two words are helpful to understanding coherence: flow and relationship. Earlier, transitions were referred to as being the "glue" to put organized thoughts together. Now, let's look at the topic sentence from which flow and relationship originate.

The topic sentence, usually the first in a paragraph, holds the essential features that will be brought forth in the paragraph. It is also here that authors either grab or lose readers. It may be the only writing that a reader encounters from that writer, so it is a good idea to summarize and represent ideas accurately.

The coherent paragraph has a logical order. It utilizes transitional words and phrases, parallel sentence structure, clear pronoun references, and reasonable repetition of key words and phrases. Use common sense for repetition. Consider synonyms for variety. Be consistent in verb tense whenever possible.

When writers have accomplished their paragraph's purpose, they prepare it to receive the next paragraph. While writing, read the paragraph over, edit, examine, evaluate, and make changes accordingly. Possibly, a paragraph has gone on too long. If that occurs, it needs to be broken up into other paragraphs, or the length should be reduced. If a paragraph didn't fully accomplish its purpose, consider revising it.

Transitions

Transitions are the glue used to make organized thoughts adhere to one another. Transitions help blend ideas together seamlessly, within sentences and paragraphs, between them, and (in longer documents) even between sections. Transitions may be single words, sentences, or whole paragraphs. Transitions help readers to digest and understand what to feel about what has gone on and clue readers in on what is going on, what will be, and how they might react to all these factors. Transitions are like good clues left at a crime scene.

Transitions have many emphases as can be seen below.

- To show emphasis: truly, in fact
- To show examples: for example, namely, specifically
- To show similarities: also, likewise
- To show dissimilarities: on the other hand, even if, in contrast
- To show progression of time: later, previously, subsequently
- To show sequence or order: next, finally
- To show cause and effect: therefore, so
- To show place or position: above, nearby, there
- To provide evidence: furthermore, then
- To summarize: finally, summarizing

Math

Numbers and Operations

Properties of Operations with Real Numbers

The mathematical number system is made up of two general types of numbers: real and complex. **Real numbers** are those that are used in normal settings, while **complex numbers** are those composed of both a real number and an imaginary one. Imaginary numbers are the result of taking the square root of -1, and $\sqrt{-1} = i$.

The real number system is often explained using a Venn diagram similar to the one below. After a number has been labeled as a real number, further classification occurs when considering the other groups in this diagram. If a number is a never-ending, non-repeating decimal, it falls in the irrational category. Otherwise, it is rational. Furthermore, if a number does not have a fractional part, it is classified as an integer, such as -2, 75, or zero. Whole numbers are an even smaller group that only includes positive integers and zero. The last group of natural numbers is made up of only positive integers, such as 2, 56, or 12.

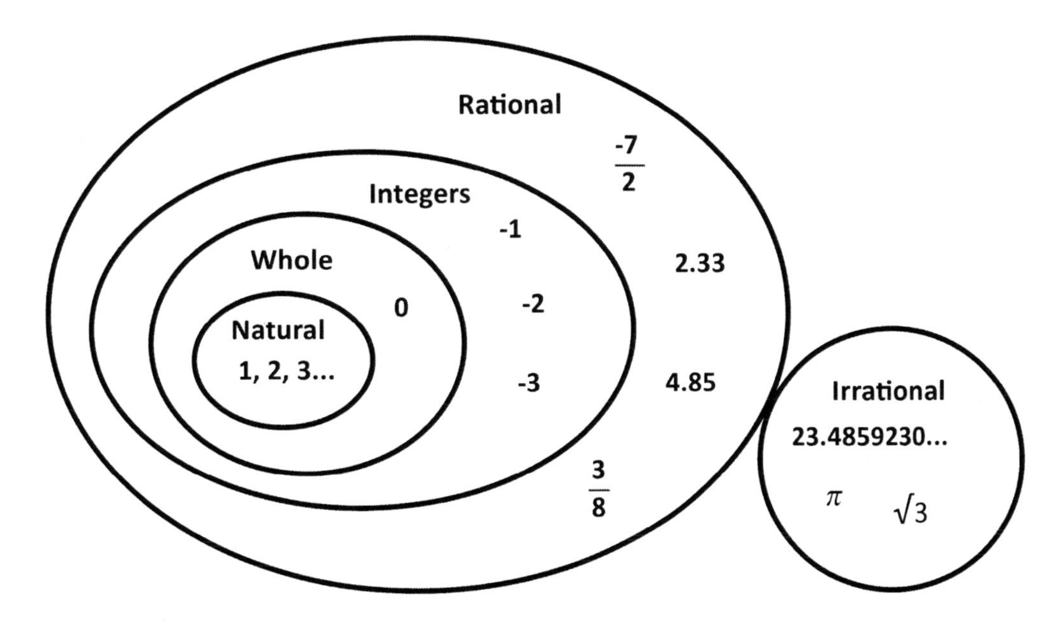

Both positive and negative numbers are valued according to their distance from 0. Both +3 and -3 can be considered using the following number line:

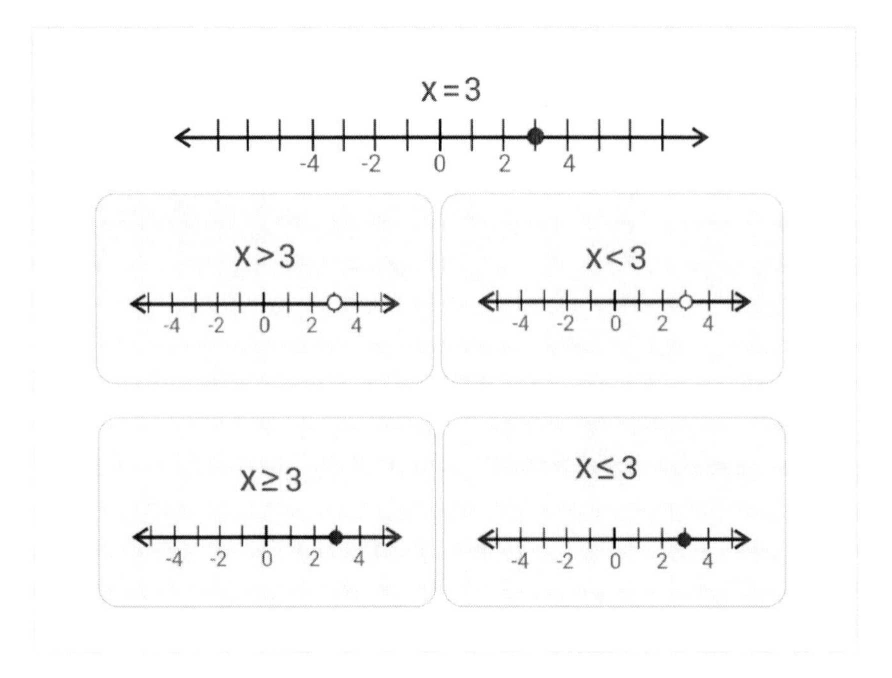

Both 3 and -3 are three spaces from 0. The distance from 0 is called its **absolute value**. Thus, both -3 and 3 have an absolute value of 3 since they're both three spaces away from 0.

An absolute number is written by placing | | around the number. So, $|3|$ and $|-3|$ both equal 3, as that's their common absolute value.

Real numbers can be compared and ordered using the number line. If a number falls to the left on the real number line, it is less than a number on the right. For example, $-2 < 5$ because -2 falls to the left of zero, and 5 falls to the right. Numbers to the left of zero are negative while those to the right are positive.

The order of operations—PEMDAS—simplifies longer expressions with real or imaginary numbers. Each operation is listed in the order of how they should be completed in a problem containing more than one operation. Parenthesis can also mean grouping symbols, such as brackets and absolute value. Then, exponents are calculated. Multiplication and division should be completed from left to right, and addition and subtraction should be completed from left to right. The following shows step-by-step how an expression is simplified using the order of operations:

$$25 \div (8 - 3)^2 - 1$$

$$25 \div (5)^2 - 1$$

$$25 \div 25 - 1$$

$$1 - 1$$

$$0$$

Simplification of another type of expression occurs when radicals are involved. For example, the following expression is a radical, or root, that can be simplified: $\sqrt{24x^2}$. First, the number must be factored out to the highest perfect square. Any perfect square can be taken out of a radical. Twenty-four can be factored into 4 and 6, and 4 can be taken out of the radical. $\sqrt{4} = 2$ can be taken out, and 6 stays underneath. If $x > 0$, x can be taken out of the radical because it is a perfect square. The simplified radical is $2x\sqrt{6}$. An approximation can be found using a calculator.

There are also properties of numbers that are true for certain operations. The **commutative** property allows the order of the terms in an expression to change while keeping the same final answer. Both addition and multiplication can be completed in any order and still obtain the same result. However, order does matter in subtraction and division. The **associative** property allows any terms to be "associated" by parenthesis and retain the same final answer. For example,

$$(4+3)+5 = 4+(3+5)$$

Both addition and multiplication are associative; however, subtraction and division do not hold this property. The **distributive** property states that $a(b+c) = ab + ac$. It is a property that involves both addition and multiplication, and the *a* is distributed onto each term inside the parentheses.

Integers can be factored into prime numbers. To **factor** is to express as a product. For example,

$$6 = 3 \times 2 \text{ and } 6 = 6 \times 1$$

Both are factorizations, but the expression involving the factors of 3 and 2 is known as a **prime factorization** because it is factored into a product of two **prime numbers**—integers which do not have any factors other than themselves and 1. A **composite number** is a positive integer that can be divided into at least one other integer other than itself and 1, such as 6. Integers that have a factor of 2 are even, and if they are not divisible by 2, they are odd. Finally, a **multiple** of a number is the product of that number and a counting number—also known as a **natural number**. For example, some multiples of 4 are 4, 8, 12, 16, etc.

The **greatest common factor** is the largest number among the shared, common factors. From the factors of 45 and 30, the common factors are 3, 5, and 15. Thus, 15 is the greatest common factor, as it's the largest number.

The **least common multiple** is the smallest number that's a multiple of two numbers. For example, to find the least common multiple of 4 and 9, the multiples of 4 and 9 are found first. The multiples of 4 are 4, 8, 12, 16, 20, 24, 28, 32, 36, and so on. For 9, the multiples are 9, 18, 27, 36, 45, 54, etc. Thus, the least common multiple of 4 and 9 is 36, the lowest number where 4 and 9 share multiples.

If two numbers share no factors besides 1 in common, then their least common multiple will be simply their product. If two numbers have common factors, then their least common multiple will be their product divided by their greatest common factor. This can be visualized by the formula $LCM = \frac{x \times y}{GCF}$, where *x* and *y* are some integers, and LCM and GCF are their least common multiple and greatest common factor, respectively.

Operations with Fractions

A **fraction** is an equation that represents a part of a whole but can also be used to present ratios or division problems. An example of a fraction is $\frac{x}{y}$. In this example, x is called the **numerator**, while y is the

denominator. The numerator represents the number of parts, and the denominator is the total number of parts. They are separated by a line or slash, known as a fraction bar. In simple fractions, the numerator and denominator can be nearly any integer. However, the denominator of a fraction can never be zero because dividing by zero is a function which is undefined.

Fractions come in three different varieties: proper fractions, improper fractions, and mixed numbers. **Proper fractions** have a numerator less than the denominator, such as $\frac{3}{8}$, but **improper fractions** have a numerator greater than the denominator, such as:

$$\frac{15}{8}$$

Mixed numbers combine a whole number with a proper fraction, such as:

$$3\frac{1}{2}$$

Any mixed number can be written as an improper fraction by multiplying the integer by the denominator, adding the product to the value of the numerator, and dividing the sum by the original denominator. For example:

$$3\frac{1}{2}$$

$$\frac{3 \times 2 + 1}{2}$$

$$\frac{7}{2}$$

Whole numbers can also be converted into fractions by placing the whole number as the numerator and making the denominator 1. For example:

$$3 = \frac{3}{1}$$

One of the most fundamental concepts of fractions is their ability to be manipulated by multiplication or division. This is possible since $\frac{n}{n} = 1$ for any non-zero integer. As a result, multiplying or dividing by $\frac{n}{n}$ will not alter the original fraction since any number multiplied or divided by 1 doesn't change the value of that number. Fractions of the same value are known as equivalent fractions. For example, $\frac{2}{4}, \frac{4}{8}, \frac{50}{100}$, and $\frac{75}{150}$ are equivalent, as they all equal $\frac{1}{2}$.

Of the four basic operations that can be performed on fractions, the one that involves the least amount of work is multiplication. To multiply two fractions, the numerators are multiplied, the denominators are multiplied, and the products are placed together as a fraction. Whole numbers and mixed numbers can also be expressed as a fraction which more easily facilitates multiplication with another fraction. The following problems provide examples:

1. $\frac{2}{5} \times \frac{3}{4} = \frac{6}{20} = \frac{3}{10}$

2. $\frac{4}{9} \times \frac{7}{11} = \frac{28}{99}$

Dividing fractions is similar to multiplication with one key difference. To divide fractions, the numerator and denominator of the second fraction are flipped, and then one proceeds as if it were a multiplication problem:

1. $\frac{7}{8} \div \frac{4}{5} = \frac{7}{8} \times \frac{5}{4} = \frac{35}{32}$
2. $\frac{5}{9} \div \frac{1}{3} = \frac{5}{9} \times \frac{3}{1} = \frac{15}{9} = \frac{5}{3}$

Addition and subtraction require more steps than multiplication and division, as these operations require the fractions to have the same denominator, also called a **common denominator**. It is always possible to find a common denominator by multiplying the denominators. However, when the denominators are large numbers, this method is unwieldy, especially if the answer must be provided in its simplest form. Thus, it's beneficial to find the least common denominator of the fractions—the least common denominator is incidentally also the least common multiple.

Once equivalent fractions have been found with common denominators, the numerators are simply added or subtracted to arrive at the answer:

1. $\frac{1}{2} + \frac{3}{4} = \frac{2}{4} + \frac{3}{4} = \frac{5}{4}$
2. $\frac{3}{12} + \frac{11}{20} = \frac{15}{60} + \frac{33}{60} = \frac{48}{60} = \frac{4}{5}$
3. $\frac{7}{9} - \frac{4}{15} = \frac{35}{45} - \frac{12}{45} = \frac{23}{45}$
4. $\frac{5}{6} - \frac{7}{18} = \frac{15}{18} - \frac{7}{18} = \frac{8}{18} = \frac{4}{9}$

Percentages

Percentages can be thought of as fractions with a denominator of 100. In fact, percentage means "per hundred." Problems often require converting numbers from percentages, fractions, and decimals. The following explains how to work through those conversions.

Converting Fractions to Percentages

The fraction is converted by using an equivalent fraction with a denominator of 100. For example:

$$\frac{3}{4}$$

$$\frac{3}{4} \times \frac{25}{25}$$

$$\frac{75}{100}$$

$$75\%$$

Converting Percentages to Fractions: Percentages can be converted to fractions by turning the percentage into a fraction with a denominator of 100. Test takers should be wary of questions asking the converted fraction to be written in the simplest form. For example, $35\% = \frac{35}{100}$ which, although correctly written, has a numerator and denominator with a greatest common factor of 5, so it can be simplified to $\frac{7}{20}$.

Converting Percentages to Decimals: Because a percentage is based on "per hundred," decimals and percentages can be converted by multiplying or dividing by 100. Practically speaking, this always amounts to moving the decimal point two places to the right or left, depending on the conversion. To convert a percentage to a decimal, the decimal point is moved two places to the left and the % sign gets removed. To convert a decimal to a percentage, the decimal point is moved two places to the right and a "%" sign is added. Here are some examples:

65% = 0.65

0.33 = 33%

0.215 = 21.5%

99.99% = 0.9999

500% = 5.00

7.55 = 755%

Properties of Rational and Irrational Numbers

All real numbers can be separated into two groups: rational and irrational numbers. **Rational numbers** are any numbers that can be written as a fraction, such as $\frac{1}{3}, \frac{7}{4}$, and -25. Alternatively, **irrational numbers** are those that cannot be written as a fraction, such as numbers with never-ending, non-repeating decimal values. Many irrational numbers result from taking roots, such as $\sqrt{2}$ or $\sqrt{3}$. An irrational number may be written as:

34.5684952...

The ellipsis (...) represents the line of numbers after the decimal that does not repeat and is never-ending.

When rational and irrational numbers interact, there are different types of number outcomes. For example, when adding or multiplying two rational numbers, the result is a rational number. No matter what two fractions are added or multiplied together, the result can always be written as a fraction. The following expression shows two rational numbers multiplied together:

$$\frac{3}{8} \times \frac{4}{7} = \frac{12}{56}$$

The product of these two fractions is another fraction that can be simplified to $\frac{3}{14}$.

As another interaction, rational numbers added to irrational numbers will always result in irrational numbers. No part of any fraction can be added to a never-ending, non-repeating decimal to make a rational number. The same result is true when multiplying a rational and irrational number. Taking a fractional part of a never-ending, non-repeating decimal will always result in another never-ending, non-repeating decimal. An example of the product of rational and irrational numbers is shown in the following expression: $2\times \sqrt{7}$.

The last type of interaction concerns two irrational numbers, where the sum or product may be rational or irrational depending on the numbers being used. The following expression shows a rational sum from two irrational numbers:

$$\sqrt{3} + \left(6 - \sqrt{3}\right) = 6$$

The product of two irrational numbers can be rational or irrational. A rational result can be seen in the following expression:

$$\sqrt{2} \times \sqrt{8} = \sqrt{2 \times 8} = \sqrt{16} = 4$$

An irrational result can be seen in the following:

$$\sqrt{3} \times \sqrt{2} = \sqrt{6}$$

Rewriting Expressions Involving Radicals and Rational Exponents

Exponents are used in mathematics to express a number or variable multiplied by itself a certain number of times. For example, x^3 means *x* is multiplied by itself three times. In this expression, *x* is called the **base**, and 3 is the **exponent**. Exponents can be used in more complex problems when they contain fractions and negative numbers.

Fractional exponents can be explained by looking first at the inverse of exponents, which are **roots**. Given the expression x^2, the square root can be taken, $\sqrt{x^2}$, cancelling out the 2 and leaving *x* by itself, if *x* is positive. Cancellation occurs because $\sqrt{x}$ can be written with exponents, instead of roots, as $x^{\frac{1}{2}}$. The numerator of 1 is the exponent, and the denominator of 2 is called the root (which is why it's referred to as square root). Taking the square root of x^2 is the same as raising it to the $\frac{1}{2}$ power. Written out in mathematical form, it takes the following progression:

$$\sqrt{x^2} = (x^2)^{\frac{1}{2}} = x$$

From properties of exponents, $2 \times \frac{1}{2} = 1$ is the actual exponent of *x*. Another example can be seen with $x^{\frac{4}{7}}$. The variable *x*, raised to four-sevenths, is equal to the seventh root of *x* to the fourth power: $\sqrt[7]{x^4}$. In general,

$$x^{\frac{1}{n}} = \sqrt[n]{x}$$

and

$$x^{\frac{m}{n}} = \sqrt[n]{x^m}$$

Negative exponents also involve fractions. Whereas y^3 can also be rewritten as $\frac{y^3}{1}$, y^{-3} can be rewritten as $\frac{1}{y^3}$. A negative exponent means the exponential expression must be moved to the opposite spot in a fraction to make the exponent positive. If the negative appears in the numerator, it moves to the denominator. If the negative appears in the denominator, it is moved to the numerator. In general, $a^{-n} = \frac{1}{a^n}$, and a^{-n} and a^n are reciprocals.

Take, for example, the following expression:

$$\frac{a^{-4}b^2}{c^{-5}}$$

Since *a* is raised to the negative fourth power, it can be moved to the denominator. Since *c* is raised to the negative fifth power, it can be moved to the numerator. The *b* variable is raised to the positive second power, so it does not move.

The simplified expression is as follows:

$$\frac{b^2c^5}{a^4}$$

In mathematical expressions containing exponents and other operations, the order of operations must be followed. *PEMDAS* states that exponents are calculated after any parenthesis and grouping symbols but before any multiplication, division, addition, and subtraction.

Scientific Notation

Scientific notation is used to represent numbers that are either very small or very large. For example, the distance to the sun is approximately 150,000,000,000 meters. Instead of writing this number with so many zeros, it can be written in scientific notation as 1.5×10^{11}meters. The same is true for very small numbers, but the exponent becomes negative. If the mass of a human cell is 0.000000000001 kilograms, that measurement can be easily represented by 1.0×10^{-12} kilograms. In both situations, scientific notation makes the measurement easier to read and understand. Each number is translated to an expression with one digit in the tens place times an expression corresponding to the zeros.

When two measurements are given and both involve scientific notation, it is important to know how these interact with each other:

- In addition and subtraction, the exponent on the ten must be the same before any operations are performed on the numbers. For example,

  $$(1.3 \times 10^4) + (3.0 \times 10^3)$$

  This cannot be added until one of the exponents on the ten is changed. The 3.0×10^3 can be changed to 0.3×10^4, then the 1.3 and 0.3 can be added. The answer comes out to be 1.6×10^4.

- For multiplication, the first numbers can be multiplied and then the exponents on the tens can be added. Once an answer is formed, it may have to be converted into scientific notation again depending on the change that occurred.

- The following is an example of multiplication with scientific notation:

$$(4.5 \times 10^3) \times (3.0 \times 10^{-5}) = 13.5 \times 10^{-2}$$

- Since this answer is not in scientific notation, the decimal is moved over to the left one unit, and 1 is added to the ten's exponent. This results in the final answer: 1.35×10^{-1}.

- For division, the first numbers are divided, and the exponents on the tens are subtracted. Again, the answer may need to be converted into scientific notation form, depending on the type of changes that occurred during the problem.

- Order of magnitude relates to scientific notation and is the total count of powers of 10 in a number. For example, there are 6 orders of magnitude in 1,000,000. If a number is raised by an order of magnitude, it is multiplied times 10. Order of magnitude can be helpful in estimating results using very large or small numbers. An answer should make sense in terms of its order of magnitude.

- For example, if area is calculated using two dimensions with 6 orders of magnitude, because area involves multiplication, the answer should have around 12 orders of magnitude. Also, answers can be estimated by rounding to the largest place value in each number. For example, 5,493,302×2,523,100 can be estimated by $5 \times 3 = 15$ with 6 orders of magnitude.

Reasoning Quantitatively

Dimensional analysis is the process of converting between different units using equivalent measurement statements. For example, running a 5K is the same as running approximately 3.1 miles. This conversion can be found by knowing that 1 kilometer is equal to approximately 0.62 miles.

The following calculation shows how to convert kilometers into miles. The original units need to be opposite one another in each of the two fractions: one in the original amount and one in the denominator of the conversion factor. This specific example consists of 5 km being multiplied times the conversion factor .62 mi/km. By design, quantities in kilometers are opposite one another and therefore cancel, leaving 3.11 miles as the converted result.

$$5\cancel{km} \times \left(\frac{0.62miles}{1\cancel{km}}\right) = 3.11\ miles$$

Units are also important throughout formulas in calculating quantities such as volume and area. To find the volume of a pyramid, the following formula is used: $V = \frac{1}{3}Bh$. *B* is the area of the base, and *h* is the height. In the example shown below, two of the same type of dimension are composed of two different units. All dimensions must be converted to the same units before plugging values into the formula for volume.

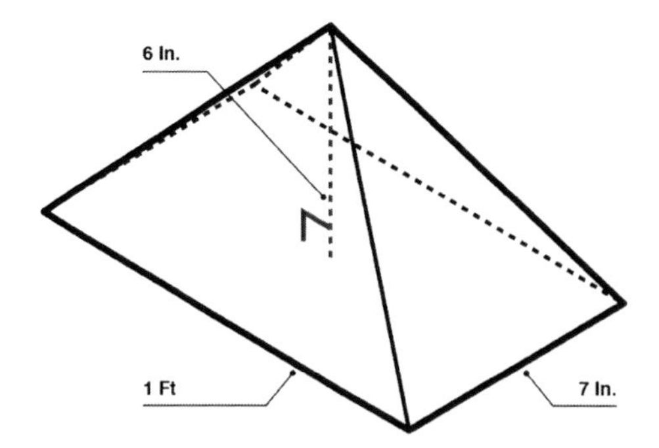

In this case, all lengths will be converted to inches. To find the area of the base, it's necessary to convert 1 ft. to 12 inches. Then, the area of the base can be calculated as:

$$B = 12\ in \times 7\ in = 84\ in^2$$

B can then be substituted into the volume formula as follows:

$$V = \frac{1}{3}(84in^2)(6in) = 168\ in^3$$

Formulas are a common situation in which units need to be interpreted and used. However, graphs can also carry meaning through units. The following graph is an example. It represents a graph of the position of an object over time. The *m* axis represents the number of meters the object is from the starting point at time *s*, in seconds. Interpreting this graph, the origin shows that at time zero seconds, the object is zero meters away from the starting point. As the time increases to one second, the position increases to five meters away. This trend continues until 6 seconds, where the object is 30 meters away from the starting position. After this point in time—since the graph remains horizontal from 6 to 10 seconds—the object must have stopped moving.

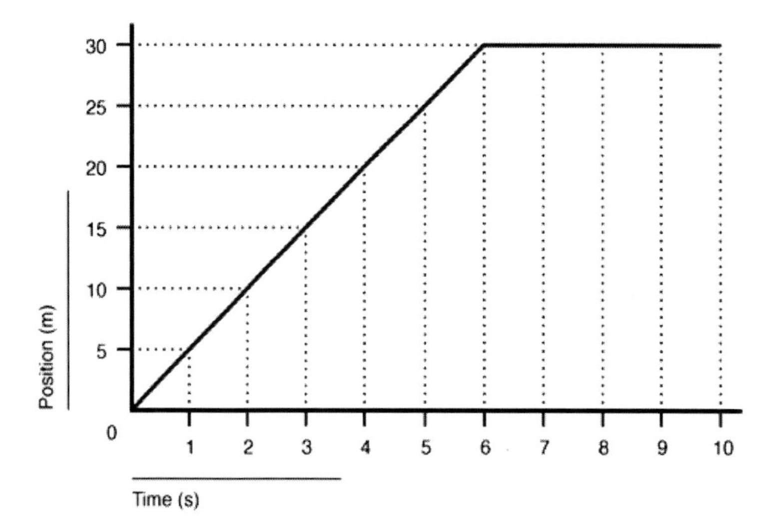

In each of the previous problem examples, the units were important to the answer. When solving problems with units, it's important to consider the reasonableness of the answer. If conversions are used, it's helpful to have an estimated value to compare the final answer to. This way, if the final answer is too distant from the estimate, it will be obvious that a mistake was made.

Choosing a Level of Accuracy

Precision and accuracy are used to describe groups of measurements. **Precision** describes a group of measures that are very close together, regardless of whether the measures are close to the true value. **Accuracy** describes how close the measures are to the true value.

The following graphic illustrates the different combinations that may occur with different groups of measures:

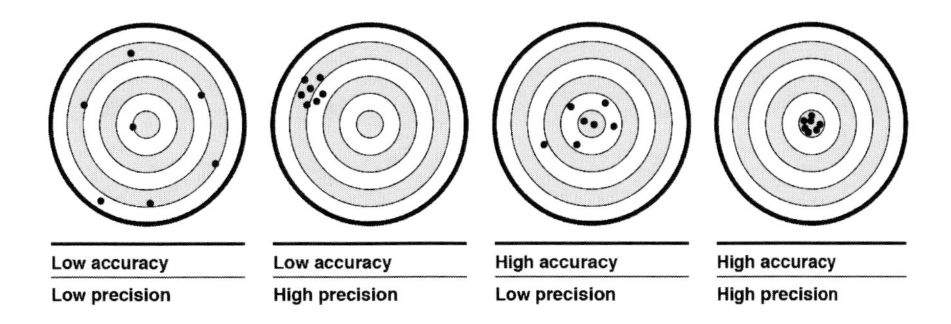

Since accuracy refers to the closeness of a value to the true measurement, the level of accuracy depends on the object measured and the instrument used to measure it. This will vary depending on the situation. If measuring the mass of a set of dictionaries, kilograms may be used as the units. In this case, it is not vitally important to have a high level of accuracy. If the measurement is a few grams away from the true value, the discrepancy might not make a big difference in the problem.

In a different situation, the level of accuracy may be more significant. Pharmacists need to be sure they are very accurate in their measurements of medicines that they give to patients. In this case, the level of accuracy is vitally important and not something to be estimated. In the dictionary situation, the measurements were given as whole numbers in kilograms. In the pharmacist's situation, the measurements for medicine must be taken to the milligram and sometimes further, depending on the type of medicine.

When considering the accuracy of measurements, the error in each measurement can be shown as absolute and relative. **Absolute error** tells the actual difference between the measured value and the true value. The **relative error** tells how large the error is in relation to the true value. There may be two problems where the absolute error of the measurements is 10 grams. For one problem, this may mean the relative error is very small because the measured value is 14,990 grams, and the true value is 15,000 grams. Ten grams in relation to the true value of 15,000 is small: 0.06%. For the other problem, the measured value is 290 grams, and the true value is 300 grams. In this case, the 10-gram absolute error means a high relative error because the true value is smaller. The relative error is 10/300 = 0.03, or 3%.

Solving Multistep Real-World Problems

Ratios are used to show the relationship between two quantities. The ratio of oranges to apples in the grocery store may be 3 to 2. That means that for every 3 oranges, there are 2 apples. This comparison can be expanded to represent the actual number of oranges and apples, such as 36 oranges to 24 apples. Another example may be the number of boys to girls in a math class. If the ratio of boys to girls is given as 2 to 5, that means there are 2 boys to every 5 girls in the class. Ratios can also be compared if the units in each ratio are the same. The ratio of boys to girls in the math class can be compared to the ratio of boys to girls in a science class by stating which ratio is higher and which is lower.

Rates are used to compare two quantities with different units. **Unit rates** are the simplest form of rate. With unit rates, the denominator in the comparison of two units is one. For example, if someone can type at a rate of 1000 words in 5 minutes, then their unit rate for typing is $\frac{1000}{5} = 200$ words in one minute or 200 words per minute. Any rate can be converted into a unit rate by dividing to make the denominator one. 1000 words in 5 minutes has been converted into the unit rate of 200 words per minute.

The ratio between two similar geometric figures is called the **scale factor**. In the following example, there are two similar triangles. The scale factor from figure A to figure B is 2 because the length of the corresponding side of the larger triangle, 14, is twice the corresponding side on the smaller triangle, 7.

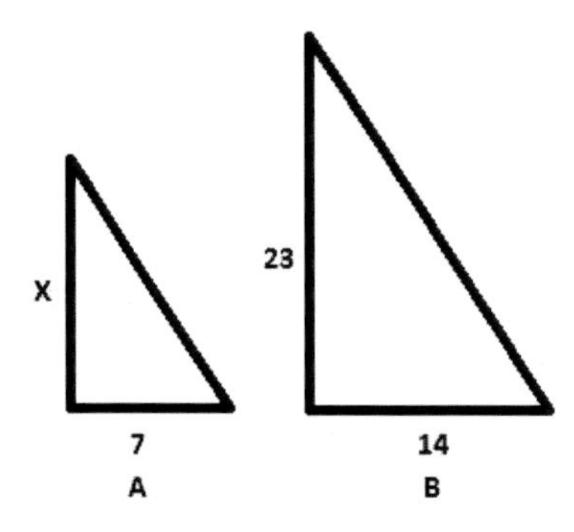

This scale factor can also be used to find the value of X. Since the scale factor from small to large is 2, the larger number, 23, can be divided by 2 to find the missing side: $X = 11.5$. The scale factor can also be represented in the equation $2A = B$ because two times the lengths of A gives the corresponding lengths of B. This is the idea behind similar triangles.

Much like a scale factor can be written using an equation like $2A = B$, a **proportional relationship** is represented by the equation $Y = kX$. X and Y are proportional because as values in X increase, the values in Y also increase. A relationship that is inversely proportional can be represented by the equation $Y = \frac{k}{X}$, where the value of Y decreases as the value of X increases and vice versa. The following graph represents these two types of relationships between x and y. The grey line represents a proportional relationship because the y-values increase as the x-values increase.

The black line represents an inversely-proportional relationship because the y-values decrease as the x-values increase.

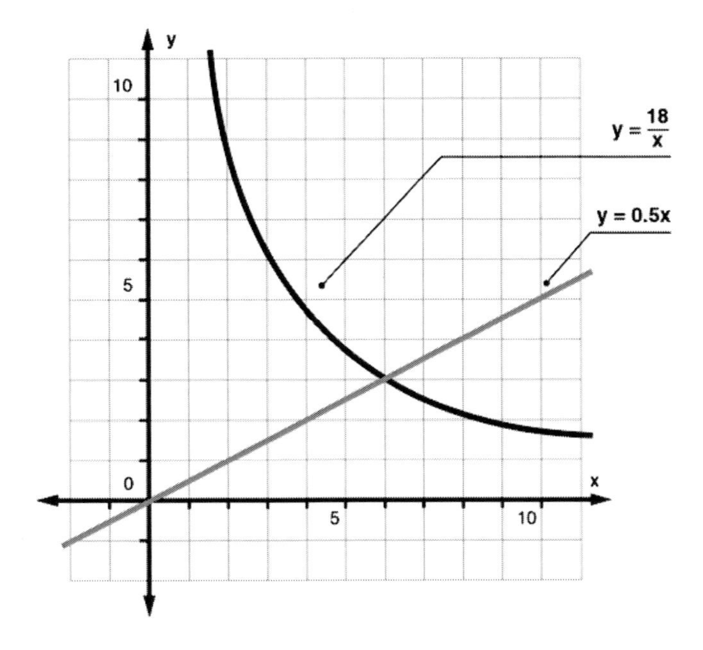

Proportional reasoning can be used to solve problems involving ratios, percentages, and averages. Ratios can be used in setting up proportions and solving them to find unknowns. For example, if someone averages 10 pages of math homework completed in 3 nights, how long would it take him or her to complete 22 pages? Both ratios can be written as fractions. The second ratio would contain the unknown. The following proportion represents this problem where x is the unknown number of nights:

$$\frac{10\ pages}{3\ nights} = \frac{22\ pages}{x\ nights}$$

Solving this proportion entails cross-multiplying and results in the following equation:

$$10x = 22 \times 3$$

Simplifying and solving for *x* results in the exact solution: $x = 6.6\ nights$. The result would be rounded up to 7 because the homework would actually be completed on the 7th night.

The following problem uses ratios involving percentages:

If 20% of the class is girls and 30 students are in the class, how many girls are in the class?

To set up this problem, it is helpful to use the common proportion:

$$\frac{\%}{100} = \frac{is}{of}$$

Within the proportion, % is the percentage of girls, 100 is the total percentage of the class, *is* is the number of girls, and *of* is the total number of students in the class. Most percentage problems can be written using this language. To solve this problem, the proportion should be set up as $\frac{20}{100} = \frac{x}{30}$, then solved for x. Cross-multiplying results in the equation,

$$20 \times 30 = 100x$$

This results in the solution $x = 6$. There are 6 girls in the class.

Problems involving volume, length, and other units can also be solved using ratios. If the following graphic of a cone is given, the problem may ask for the volume to be found.

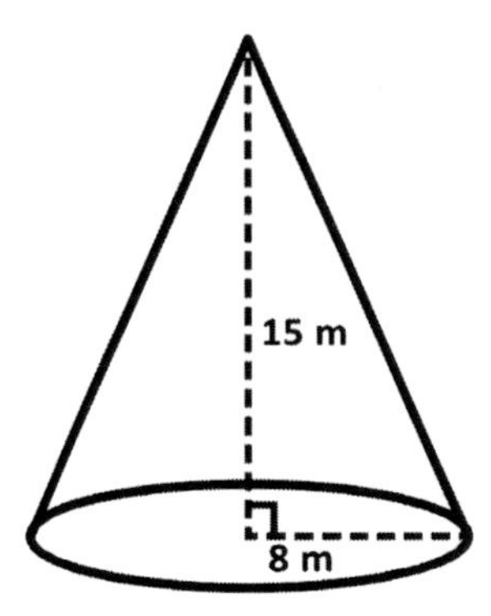

Referring to the formulas provided on the test, the volume of a cone is given as: $V = \pi r^2 \frac{h}{3}$, where r is the radius, and h is the height. Plugging $r = 8$ and $h = 15$ from the graphic into the formula, the following is obtained:

$$V = \pi(8^2)\frac{15}{3}$$

Therefore, the volume of the cone is found to be 1005.3m^3. Sometimes, answers in different units are sought. If this problem wanted the answer in liters, 1005.3m^3 would need to be converted. Using the equivalence statement 1m^3 = 1000L, the following ratio would be used to solve for liters:

$$1005.3\text{m}^3 \times \frac{1000L}{1m^3}$$

Cubic meters in the numerator and denominator cancel each other out, and the answer is converted to 1,005,300 liters, or 1.0053×10^6 L.

Algebra

Interpreting Parts of an Expression

Algebraic expressions are built out of monomials. A **monomial** is a variable raised to some power multiplied by a constant: ax^n, where *a* is any constant and *n* is a whole number. A constant is also a monomial.

A **polynomial** is a sum of monomials. Examples of polynomials include:

$$3x^4 + 2x^2 - x - 3$$

$$\frac{4}{5}x^3$$

The latter is also a monomial. If the highest power of *x* is 1, the polynomial is called **linear**. If the highest power of *x* is 2, it is called **quadratic**.

Algebraic Functions

A function is called **algebraic** if it is built up from polynomials by adding, subtracting, multiplying, dividing, and taking radicals. This means that, for example, the variable can never appear in an exponent. Thus, polynomials and rational functions are algebraic, but exponential functions are not algebraic. It turns out that logarithms and trigonometric functions are not algebraic either.

A function of the form:

$$f(x) = a_n x^n + a_{n-1}x^{n-1} + a_{n-2}x^{n-2} + \cdots + a_1 x + a_0$$

is called a polynomial function. The value of *n* is called the degree of the polynomial. In the case where $n = 1$, it is called a linear function. In the case where $n = 2$, it is called a quadratic function. In the case where $n = 3$, it is called a cubic function.

When *n* is even, the polynomial is called even, and not all real numbers will be in its range. When *n* is odd, the polynomial is called odd, and the range includes all real numbers.

The graph of a quadratic function below will be a parabola:

$$f(x) = ax^2 + bx + c$$

To see whether or not the parabola opens up or down, it's necessary to check the coefficient of x^2, which is the value a.

If the coefficient is positive, then the parabola opens upward. If the coefficient is negative, then the parabola opens downward.

Exponential Functions

An **exponential function** is a function of the form $f(x) = b^x$, where *b* is a positive real number other than 1. In such a function, *b* is called the **base**.

The **domain** of an exponential function is all real numbers, and the **range** is all positive real numbers. There will always be a horizontal asymptote of $y = 0$ on one side. If *b* is greater than 1, then the graph will be increasing moving to the right. If *b* is less than 1, then the graph will be decreasing moving to the right. Exponential functions are one-to-one. The basic exponential function graph will go through the point (0,1).

Example
Solve $5^{x+1} = 25$.

Get the x out of the exponent by rewriting the equation $5^{x+1} = 5^2$ so that both sides have a base of 5.

Since the bases are the same, the exponents must be equal to each other.

This leaves $x + 1 = 2$ or $x = 1$.

To check the answer, the x-value of 1 can be substituted back into the original equation.

Performing Arithmetic Operations on Polynomials and Rational Expressions

Addition and subtraction operations can be performed on polynomials with like terms. **Like terms** refers to terms that have the same variable and exponent. The two following polynomials can be added together by collecting like terms:

$$(x^2 + 3x - 4) + (4x^2 - 7x + 8)$$

The x^2 terms can be added as:

$$x^2 + 4x^2 = 5x^2$$

The x terms can be added as $3x + -7x = -4x$, and the constants can be added as:

$$-4 + 8 = 4$$

The following expression is the result of the addition:

$$5x^2 - 4x + 4$$

When subtracting polynomials, the same steps are followed, only subtracting like terms together.

Multiplication of polynomials can also be performed. Given the two polynomials, $(y^3 - 4)$ and $(x^2 + 8x - 7)$, each term in the first polynomial must be multiplied by each term in the second polynomial. The steps to multiply each term in the given example are as follows:

$$(y^3 \times x^2) + (y^3 \times 8x) + (y^3 \times -7) + (-4 \times x^2) + (-4 \times 8x) + (-4 \times -7)$$

Simplifying each multiplied part, yields:

$$x^2y^3 + 8xy^3 - 7y^3 - 4x^2 - 32x + 28$$

None of the terms can be combined because there are no like terms in the final expression. Any polynomials can be multiplied by each other by following the same set of steps, then collecting like terms at the end.

Polynomial Identities

Difference of squares refers to a binomial composed of the difference of two squares. For example, $a^2 - b^2$ is a difference of squares. It can be written $(a)^2 - (b)^2$, and it can be factored into $(a - b)(a + b)$. Recognizing the difference of squares allows the expression to be rewritten easily because of the form it takes. For some expressions, factoring consists of more than one step. When factoring, it's

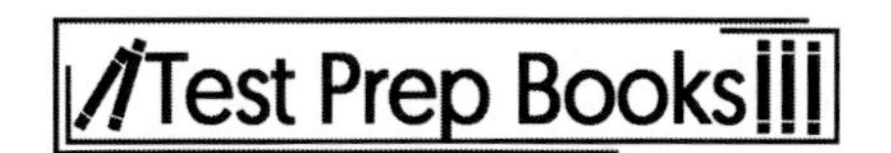

important to always check to make sure that the result cannot be factored further. If it can, then the expression should be split further. If it cannot be, the factoring step is complete, and the expression is completely factored.

A sum and difference of cubes is another way to factor a polynomial expression. When the polynomial takes the form of addition or subtraction of two terms that can be written as a cube, a formula is given. The following graphic shows the factorization of a difference of cubes:

$$a^3 - b^3 = (a - b)(a^2 + ab + b^2)$$

same sign

opposite sign

always +

This form of factoring can be useful in finding the zeros of a function of degree 3. For example, when solving $x^3 - 27 = 0$, this rule needs to be used. $x^3 - 27$ is first written as the difference two cubes, $(x)^3 - (3)^3$ and then factored into:

$$(x - 3)(x^2 + 3x + 9)$$

This expression may not be factored any further. Each factor is then set equal to zero. Therefore, one solution is found to be $x = 3$, and the other two solutions must be found using the quadratic formula. A sum of squares would have a similar process. The formula for factoring a sum of cubes is:

$$a^3 + b^3 = (a + b)(a^2 - ab + b^2)$$

The opposite of factoring is multiplying. Multiplying a square of a binomial involves the following rules:

$$(a + b)^2 = a^2 + 2ab + b^2$$

$$(a - b)^2 = a^2 - 2ab + b^2$$

Rational Expressions

A fraction, or ratio, wherein each part is a polynomial, defines **rational expressions**. Some examples include:

$$\frac{2x + 6}{x}$$

$$\frac{1}{x^2 - 4x + 8}$$

$$\frac{z^2}{x + 5}$$

Exponents on the variables are restricted to whole numbers, which means roots and negative exponents are not included in rational expressions.

Rational expressions can be transformed by factoring. For example, the expression $\frac{x^2-5x+6}{(x-3)}$ can be rewritten by factoring the numerator to obtain:

$$\frac{(x-3)(x-2)}{(x-3)}$$

Therefore, the common binomial $(x-3)$ can cancel so that the simplified expression is:

$$\frac{(x-2)}{1} = (x-2)$$

Additionally, other rational expressions can be rewritten to take on different forms. Some may be factorable in themselves, while others can be transformed through arithmetic operations. Rational expressions are closed under addition, subtraction, multiplication, and division by a nonzero expression. **Closed** means that if any one of these operations is performed on a rational expression, the result will still be a rational expression. The set of all real numbers is another example of a set closed under all four operations.

Adding and subtracting rational expressions is based on the same concepts as adding and subtracting simple fractions. For both concepts, the denominators must be the same for the operation to take place.

For example, here are two rational expressions:

$$\frac{x^3-4}{(x-3)} + \frac{x+8}{(x-3)}$$

Since the denominators are both $(x-3)$, the numerators can be combined by collecting like terms to form:

$$\frac{x^3+x+4}{(x-3)}$$

If the denominators are different, they need to be made common (the same) by using the **Least Common Denominator (LCD)**. Each denominator needs to be factored, and the LCD contains each factor that appears in any one denominator the greatest number of times it appears in any denominator. The original expressions need to be multiplied times a form of 1, which will turn each denominator into the LCD. This process is like adding fractions with unlike denominators. It is also important when working with rational expressions to define what value of the variable makes the denominator zero. For this particular value, the expression is undefined.

Multiplication of rational expressions is performed like multiplication of fractions. The numerators are multiplied; then, the denominators are multiplied. The final fraction is then simplified. The expressions

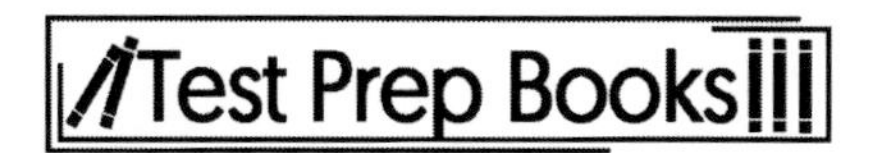

are simplified by factoring and cancelling out common terms. In the following example, the numerator of the second expression can be factored first to simplify the expression before multiplying:

$$\frac{x^2}{(x-4)} \times \frac{x^2 - x - 12}{2}$$

$$\frac{x^2}{(x-4)} \times \frac{(x-4)(x+3)}{2}$$

The $(x-4)$ on the top and bottom cancel out:

$$\frac{x^2}{1} \times \frac{(x+3)}{2}$$

Then multiplication is performed, resulting in:

$$\frac{x^3 + 3x^2}{2}$$

Dividing rational expressions is similar to the division of fractions, where division turns into multiplying by a reciprocal. The following expression can be rewritten as a multiplication problem:

$$\frac{x^2 - 3x + 7}{x - 4} \div \frac{x^2 - 5x + 3}{x - 4}$$

$$\frac{x^2 - 3x + 7}{x - 4} \times \frac{x - 4}{x^2 - 5x + 3}$$

The $x - 4$ cancels out, leaving:

$$\frac{x^2 - 3x + 7}{x^2 - 5x + 3}$$

The final answers should always be completely simplified. If a function is composed of a rational expression, the zeros of the graph can be found from setting the polynomial in the numerator as equal to zero and solving. The values that make the denominator equal to zero will either exist on the graph as a hole or a vertical asymptote.

Writing Expressions in Equivalent Forms

Algebraic expressions are made up of numbers, variables, and combinations of the two, using mathematical operations. Expressions can be rewritten based on their factors. For example, the expression $6x + 4$ can be rewritten as $2(3x + 2)$ because 2 is a factor of both $6x$ and 4. More complex expressions can also be rewritten based on their factors. The expression $x^4 - 16$ can be rewritten as,

$$(x^2 - 4)(x^2 + 4)$$

This is a different type of factoring, where a difference of squares is factored into a sum and difference of the same two terms. With some expressions, the factoring process is simple and only leads to a different way to represent the expression. With others, factoring and rewriting the expression leads to more information about the given problem.

In the following quadratic equation, factoring the binomial leads to finding the zeros of the function:

$$x^2 - 5x + 6 = y$$

This equations factors into:

$$(x - 3)(x - 2) = y$$

2 and 3 are found to be the zeros of the function when y is set equal to zero. The zeros of any function are the x-values where the graph of the function on the coordinate plane crosses the x-axis.

Exponential expressions can also be rewritten, just as quadratic equations. Properties of exponents must be understood. Multiplying two exponential expressions with the same base involves adding the exponents:

$$a^m a^n = a^{m+n}$$

Dividing two exponential expressions with the same base involves subtracting the exponents:

$$\frac{a^m}{a^n} = a^{m-n}$$

Raising an exponential expression to another exponent includes multiplying the exponents:

$$(a^m)^n = a^{mn}$$

The zero power always gives a value of 1:

$$a^0 = 1$$

Raising either a product or a fraction to a power involves distributing that power:

$$(ab)^m = a^m b^m \text{ and } \left(\frac{a}{b}\right)^m = \frac{a^m}{b^m}$$

Finally, raising a number to a negative exponent is equivalent to the reciprocal including the positive exponent:

$$a^{-m} = \frac{1}{a^m}$$

Finding the Zeros of a Function

The zeros of a function are the points where its graph crosses the x-axis. At these points, $y = 0$. One way to find the zeros is to analyze the graph. If given the graph, the x-coordinates can be found where the line crosses the x-axis. Another way to find the zeros is to set $y = 0$ in the equation and solve for x. Depending on the type of equation, this could be done by using opposite operations, by factoring the equation, by completing the square, or by using the quadratic formula. If a graph does not cross the x-axis, then the function may have complex roots.

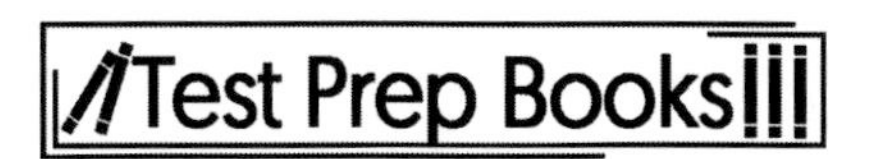

Solving Linear Equations and Inequalities

The sum of a number and 5 is equal to -8 times the number. To find this unknown number, a simple equation can be written to represent the problem. Key words such as difference, equal, and times are used to form the following equation with one variable:

$$n + 5 = -8n$$

When solving for n, opposite operations are used. First, n is subtracted from $-8n$ across the equals sign, resulting in $5 = -9n$. Then, -9 is divided on both sides, leaving $n = -\frac{5}{9}$. This solution can be graphed on the number line with a dot as shown below:

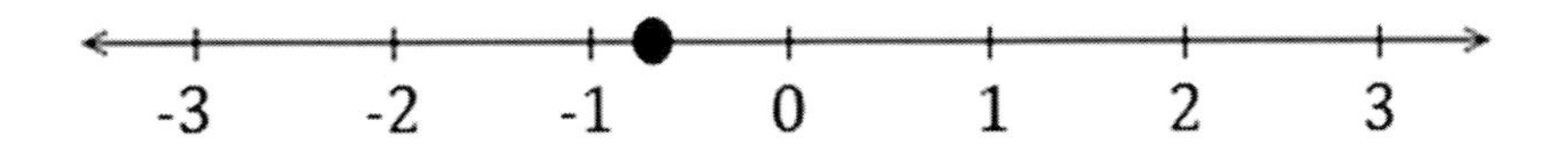

If the problem were changed to say, "The sum of a number and 5 is greater than -8 times the number," then an inequality would be used instead of an equation. Using key words again, *greater than* is represented by the symbol >. The inequality $n + 5 > -8n$ can be solved using the same techniques, resulting in $n < -\frac{5}{9}$. The only time solving an inequality differs from solving an equation is when a negative number is either multiplied times or divided by each side of the inequality. The sign must be switched in this case. For this example, the graph of the solution changes to the following graph because the solution represents all real numbers less than $-\frac{5}{9}$. Not included in this solution is $-\frac{5}{9}$ because it is a *less than* symbol, not *equal to*.

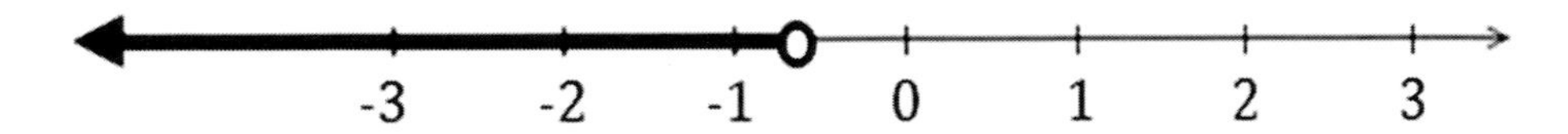

Equations and inequalities in two variables represent a relationship. Jim owns a car wash and charges $40 per car. The rent for the facility is $350 per month. An equation can be written to relate the number of cars Jim cleans to the money he makes per month. Let x represent the number of cars and y represent the profit Jim makes each month from the car wash. The equation $y = 40x - 350$ can be used to show Jim's profit or loss. Since this equation has two variables, the coordinate plane can be used to show the relationship and predict profit or loss for Jim. The following graph shows that Jim must wash at least nine cars to pay the rent, where $x = 9$. Anything nine cars and above yield a profit shown in the value on the y-axis.

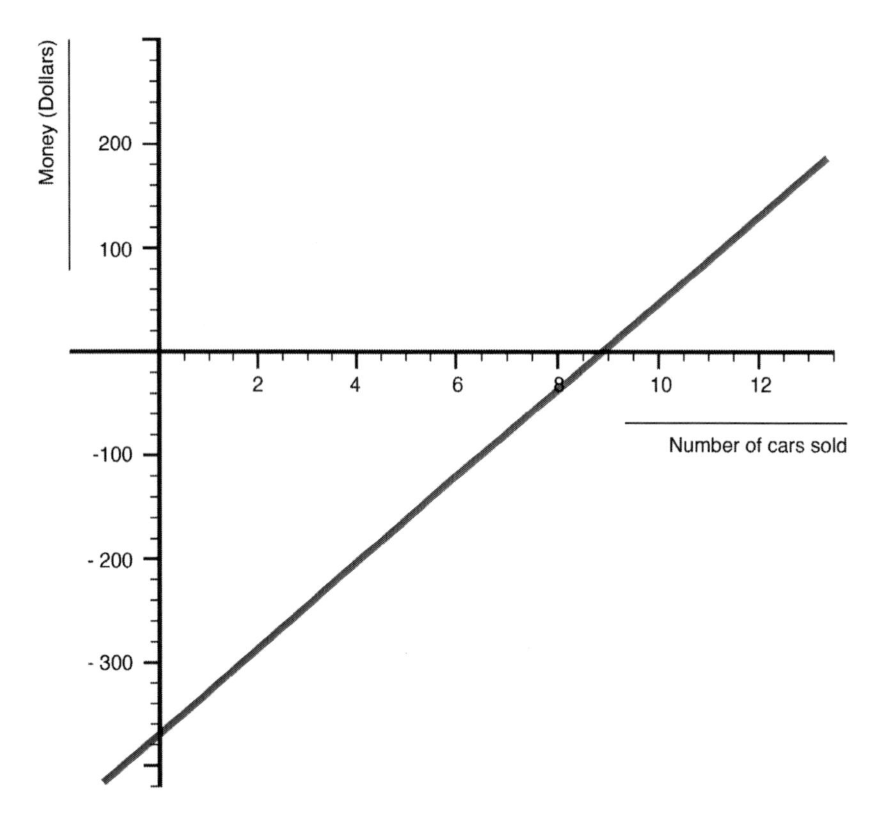

With a single equation in two variables, the solutions are limited only by the situation the equation represents. When two equations or inequalities are used, more constraints are added. For example, in a system of linear equations, there is often—although not always—only one answer. The point of intersection of two lines is the solution. For a system of inequalities, there are infinitely many answers. The intersection of two solution sets gives the solution set of the system of inequalities.

Given a linear inequality, if the inequality symbol is replaced by an equals sign, this line can be graphed. Then, if the original symbol was $<$ or $>$, either one side or the other represents all ordered pairs that satisfy this inequality. If the original symbol was $\leq$ or $\geq$, then either one side or the other plus the line represents all ordered pairs that satisfy this inequality.

Therefore, in order to graph a linear inequality, the line must be plotted first. If $<$ or $>$ is used, a dashed line is used to represent the line. If $\leq$ or $\geq$ is used, then a solid line is used to represent the line. Once the line is graphed, a test point is chosen on either side of the line. If, when plugged into the original inequality, the process results in a true statement, that side of the line is shaded in. If, when plugged into the original inequality, the process results in a false statement, the other side of the line is shaded in. Here is an example of a graph of a linear inequality:

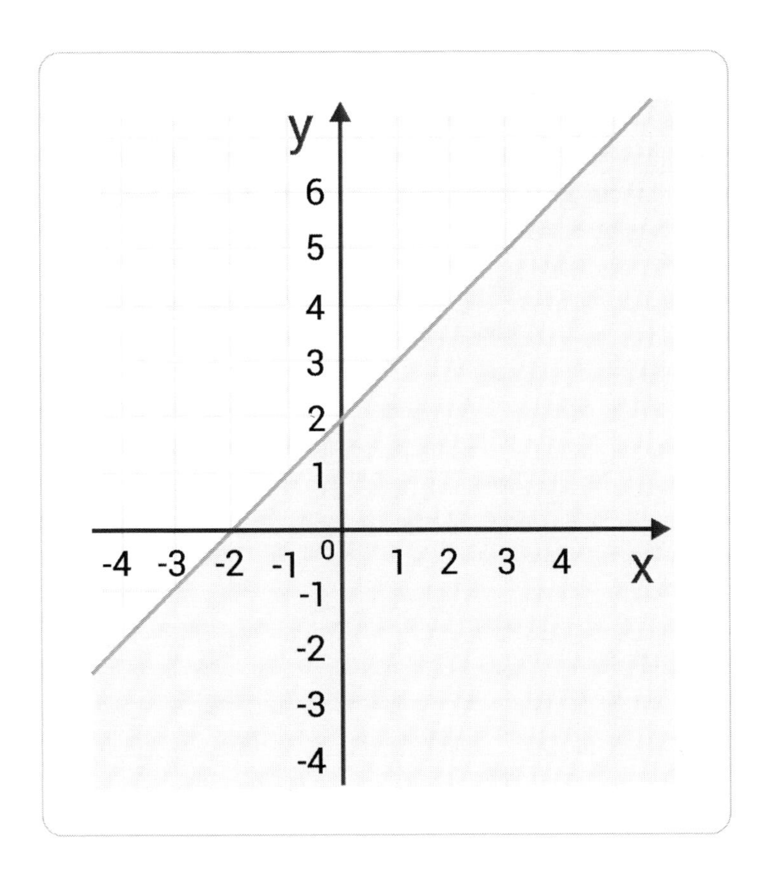

Formulas with two variables are equations used to represent a specific relationship. For example, the formula $d = rt$ represents the relationship between distance, rate, and time. If Bob travels at a rate of 35 miles per hour on his road trip from Westminster to Seneca, the formula $d = 35t$ can be used to represent his distance traveled in a specific length of time. Formulas can also be used to show different roles of the variables, transformed without any given numbers. Solving for r, the formula becomes $\frac{d}{t} = r$. The t is moved over by division so that *rate* is a function of distance and time.

Solving Equations

The letters in an equation are variables as they stand for unknown quantities that you are trying to solve for. The numbers attached to the variables by multiplication are called coefficients. *X* is commonly used as a variable, though any letter can be used. For example, in $3x - 7 = 20$, the variable is $3x$, and it needs to be isolated. The numbers (also called constants) are -7 and 20. That means $3x$ needs to be on one side of the equals sign (either side is fine), and all the numbers need to be on the other side of the equals sign.

To accomplish this, the equation must be manipulated by performing opposite operations of what already exists. Remember that addition and subtraction are opposites and that multiplication and division are opposites. Any action taken to one side of the equation must be taken on the other side to maintain equality.

Therefore, since the 7 is being subtracted, it can be moved to the right side of the equation by adding seven to both sides:

$$3x - 7 = 20$$

$$3x - 7 + 7 = 20 + 7$$

$$3x = 27$$

Now that the variable $3x$ is on one side and the constants (now combined into one constant) are on the other side, the 3 needs to be moved to the right side. 3 and x are being multiplied together, so 3 needs to be divided from each side.

$$\frac{3x}{3} = \frac{27}{3}$$

$$x = 9$$

Now that x has been completely isolated, we know its value.

The solution is found to be $x = 9$. This solution can be checked for accuracy by plugging $x = 9$ in the original equation. After simplifying the equation, $20 = 20$ is found, which is a true statement:

$$3 \times 9 - 7 = 20$$

$$27 - 7 = 20$$

$$20 = 20$$

Equations that require solving for a variable (algebraic equations) come in many forms. Here are some more examples:

No coefficient attached to the variable:

$$x + 8 = 20$$

$$x + 8 - 8 = 20 - 8$$

$$x = 12$$

A fractional coefficient:

$$\frac{1}{2}z + 24 = 36$$

$$\frac{1}{2}z + 24 - 24 = 36 - 24$$

$$\frac{1}{2}z = 12$$

Now we multiply the fraction by its inverse:

$$\frac{2}{1} \times \frac{1}{2}z = 12 \times \frac{2}{1}$$

$$z = 24$$

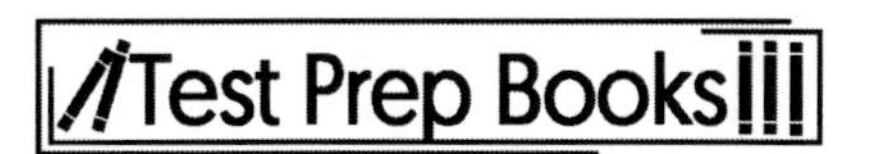

Multiple examples of x:

$$14x + x - 4 = 3x + 2$$

All examples of x can be combined.

$$15x - 4 = 3x + 2$$

$$15x - 4 + 4 = 3x + 2 + 4$$

$$15x = 3x + 6$$

$$15x - 3x = 3x + 6 - 3x$$

$$12x = 6$$

$$\frac{12x}{12} = \frac{6}{12}$$

$$x = \frac{1}{2}$$

Methods for Solving Equations

Equations with one variable can be solved using the addition principle and multiplication principle. If $a = b$, then $a + c = b + c$, and $ac = bc$. Given the equation,

$$2x - 3 = 5x + 7$$

The first step is to combine the variable terms and the constant terms. Using the principles, expressions can be added and subtracted onto and off both sides of the equals sign, so the equation turns into $-10 = 3x$. Dividing by 3 on both sides through the multiplication principle with $c = \frac{1}{3}$ results in the final answer of $x = \frac{-10}{3}$.

Solving quadratic equations is a little trickier. If they take the form $ax^2 - b = 0$, then the equation can be solved by adding *b* to both sides and dividing by *a* to get:

$$x^2 = \frac{b}{a}$$

The solution is $x = \pm\sqrt{\frac{b}{a}}$. Note that this is actually two separate solutions, unless *b* happens to be zero.

If a quadratic equation has no constant—so that it takes the form $ax^2 + bx = 0$ — then the *x* can be factored out to get:

$$x(ax + b) = 0$$

Then, the solutions are $x = 0$, together with the solutions to $ax + b = 0$. Both factors x and $(ax + b)$ can be set equal to zero to solve for x because one of those values must be zero for their product to equal zero. For an equation $ab = 0$ to be true, either $a = 0$, or $b = 0$.

A given quadratic equation $x^2 + bx + c$ can be factored into:

$$(x + A)(x + B)$$

$A + B = b$, and $AB = c$. Finding the values of *A* and *B* can take time, but such a pair of numbers can be found by guessing and checking. Looking at the positive and negative factors for *c* offers a good starting point.

For example, in $x^2 - 5x + 6$, the factors of 6 are 1, 2, and 3. Now, $(-2)(-3) = 6$, and $-2 - 3 = -5$. In general, however, this may not work, in which case another approach may need to be used.

A quadratic equation of the form $x^2 + 2xb + b^2 = 0$ can be factored into:

$$(x + b)^2 = 0$$

Similarly, $x^2 - 2xy + y^2 = 0$ factors into:

$$(x - y)^2 = 0$$

In general, the constant term may not be the right value to be factored this way. A more general method for solving these quadratic equations must then be found. The following two methods will work in any situation.

Completing the Square

Completing the square is one way to find zeros when factoring is not an option. The following equation cannot be factored:

$$x^2 + 10x - 9 = 0$$

The first step in this method is to move the constant to the right side of the equation, making it:

$$x^2 + 10x = 9$$

Then, the coefficient of x is divided by 2 and squared. This number is then added to both sides of the equation, to make the equation still true. For this example, $\left(\frac{10}{2}\right)^2 = 25$ is added to both sides of the equation to obtain:

$$x^2 + 10x + 25 = 9 + 25$$

This expression simplifies to $x^2 + 10x + 25 = 34$, which can then be factored into:

$$(x + 5)^2 = 34$$

Solving for *x* then involves taking the square root of both sides and subtracting 5. This leads to two zeros of the function:

$$x = \pm\sqrt{34} - 5$$

Depending on the type of answer the question seeks, a calculator may be used to find exact numbers.

The Quadratic Formula

The quadratic formula can be used to solve any quadratic equation. This formula may be the longest method for solving quadratic equations and is commonly used as a last resort after other methods are ruled out.

The quadratic formula is presented below:

$$x = \frac{-b \pm \sqrt{b^2 - 4ac}}{2a}$$

$a, b,$ and c are the coefficients in the original equation in standard form:

$$y = ax^2 + bx + c$$

For this equation,

$$y = x^2 - 4x + 3$$

$$x = \frac{4 \pm \sqrt{(-4)^2 - 4(1)(3)}}{2(1)}$$

$$\frac{4 \pm \sqrt{16 - 12}}{2}$$

$$\frac{4 \pm 2}{2}$$

$$1, 3$$

The expression underneath the radical is called the **discriminant**. Without working out the entire formula, the value of the discriminant can reveal the nature of the solutions. If the value of the discriminant $b^2 - 4ac$ is positive, then there will be two real solutions. If the value is zero, there will be one real solution. If the value is negative, the two solutions will be imaginary or complex. If the solutions are complex, it means that the parabola never touches the x-axis. An example of a complex solution can be found by solving the following quadratic:

$$y = x^2 - 4x + 8$$

By using the quadratic formula, the solutions are found to be:

$$x = \frac{4 \pm \sqrt{(-4)^2 - 4(1)(8)}}{2(1)}$$

$$\frac{4 \pm \sqrt{16 - 32}}{2}$$

$$\frac{4 \pm \sqrt{-16}}{2}$$

$$2 \pm 2i$$

The solutions both have a real part, 2, and an imaginary part, $2i$.

Solving Systems of Equations

A **system of equations** is a group of equations that have the same variables or unknowns. These equations can be linear, but they are not always so. Finding a solution to a system of equations means finding the values of the variables that satisfy each equation. For a linear system of two equations and two variables, there could be a single solution, no solution, or infinitely many solutions.

A single solution occurs when there is one value for x and y that satisfies the system. This is shown on the graph where the lines cross at exactly one point. When there is no solution, the lines are parallel and do not ever cross. With infinitely many solutions, the equations may look different, but they are the same line. One equation will be a multiple of the other, and on the graph, they lie on top of each other.

These three types of systems of linear equations are shown below:

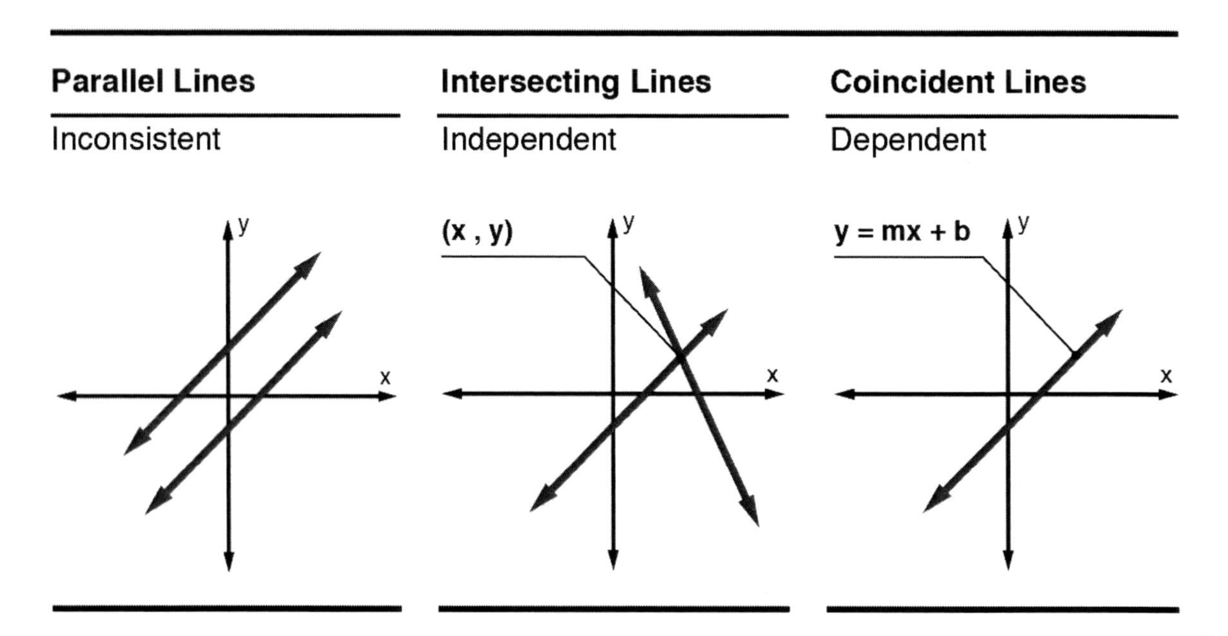

The process of elimination can be used to solve a system of equations. For example, the following equations make up a system:

$$x + 3y = 10 \text{ and } 2x - 5y = 9$$

Immediately adding these equations does not eliminate a variable, but it is possible to change the first equation by multiplying the whole equation by -2. This changes the first equation to:

$$-2x - 6y = -20$$

The equations can be then added to obtain $-11y = -11$. Solving for y yields $y = 1$. To find the rest of the solution, 1 can be substituted in for y in either original equation to find the value of $x = 7$. The solution to the system is (7, 1) because it makes both equations true, and it is the point in which the lines intersect. If the system is **dependent**—having infinitely many solutions—then both variables will cancel out when the elimination method is used, resulting in an equation that is true for many values of x and y. Since the system is dependent, both equations can be simplified to the same equation, or line.

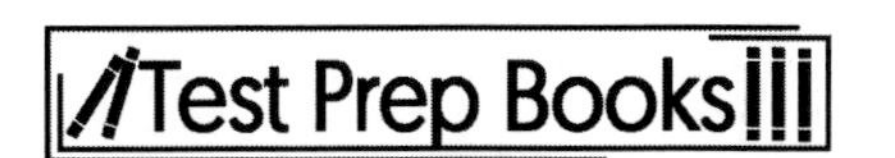

A system can also be solved using **substitution**. This involves solving one equation for a variable and then plugging that solved equation into the other equation in the system. For example, $x - y = -2$ and $3x + 2y = 9$ can be solved using substitution. The first equation can be solved for x, where:

$$x = -2 + y$$

Then it can be plugged into the other equation,

$$3(-2 + y) + 2y = 9$$

Solving for y yields $-6 + 3y + 2y = 9$, where $y = 3$. If $y = 3$, then $x = 1$. This solution can be checked by plugging in these values for the variables in each equation to see if it makes a true statement.

Finally, a solution to a system of equations can be found graphically. The solution to a linear system is the point or points where the lines cross. The values of x and y represent the coordinates (x, y) where the lines intersect. Using the same system of equations as above, they can be solved for y to put them in slope-intercept form,

$$y = mx + b$$

These equations become $y = x + 2$ and,

$$y = -\frac{3}{2}x + 4.5$$

The slope is the coefficient of x, and the y-intercept is the constant value. This system with the solution is shown below:

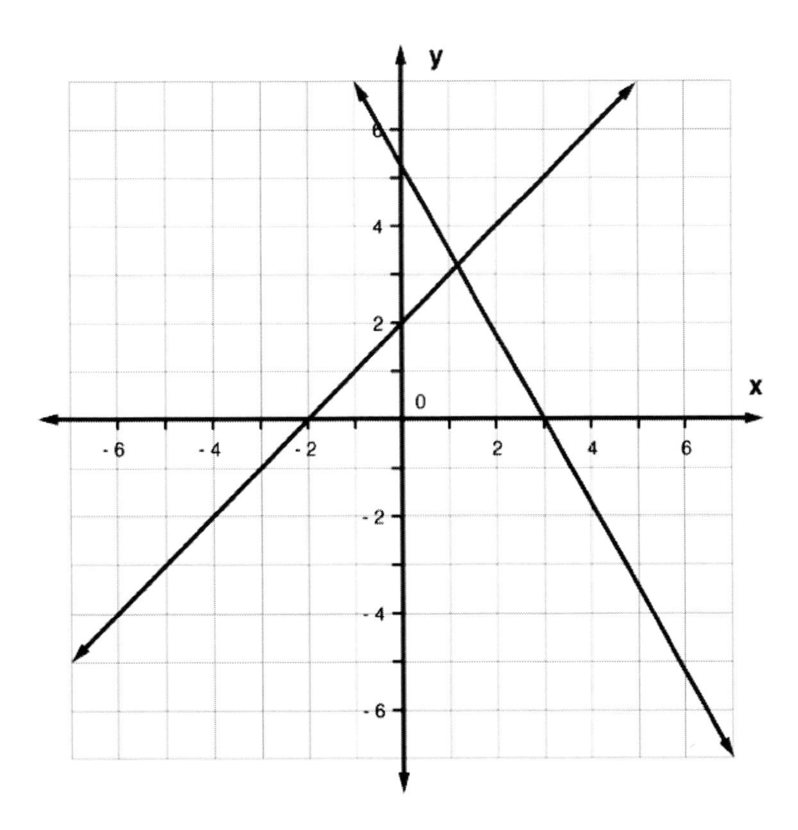

A system of equations may also be made up of a linear and a quadratic equation. These systems may have one solution, two solutions, or no solutions. The graph of these systems involves one straight line and one parabola. Algebraically, these systems can be solved by solving the linear equation for one variable and plugging that answer in to the quadratic equation. If possible, the equation can then be solved to find part of the answer. The graphing method is commonly used for these types of systems. On a graph, these two lines can be found to intersect at one point, at two points across the parabola, or at no points.

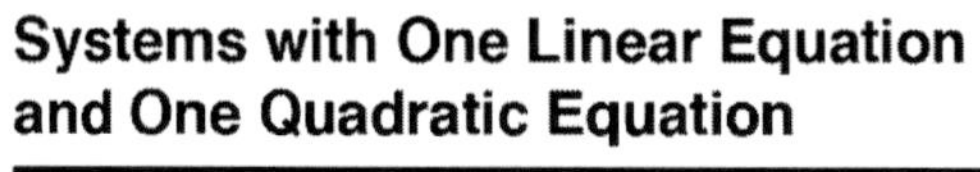

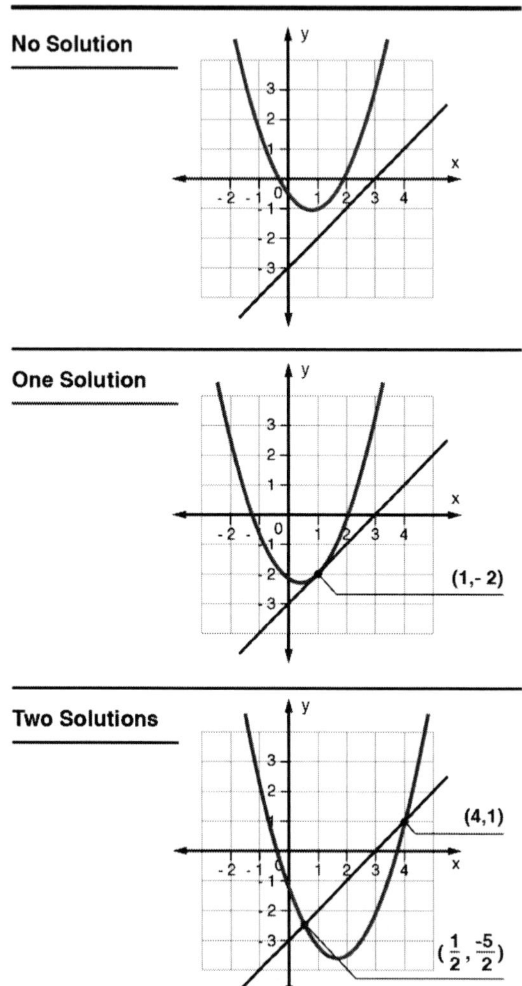

Representing and Solving Equations and Inequalities Graphically

Systems of linear inequalities are like systems of equations, but the solutions are different. Since inequalities have infinitely many solutions, their systems also have infinitely many solutions. Finding the solutions of inequalities involves graphs. A system of two equations and two inequalities is linear; thus, the lines can be graphed using slope-intercept form. If the inequality has an equals sign, the line is solid. If the inequality only has a greater than or less than symbol, the line on the graph is dotted. Dashed lines indicate that points lying on the line are not included in the solution. After the lines are graphed, a

region is shaded on one side of the line. This side is found by determining if a point—known as a *test point*—lying on one side of the line produces a true inequality. If it does, that side of the graph is shaded. If the point produces a false inequality, the line is shaded on the opposite side from the point. The graph of a system of inequalities involves shading the intersection of the two shaded regions. An example of a system of linear inequalities is shown below. The smaller shaded region that overlaps both larger shaded regions is the solution to the system. Any point that lies in this region or on the solid line in that region produces a true inequality for either inequality used.

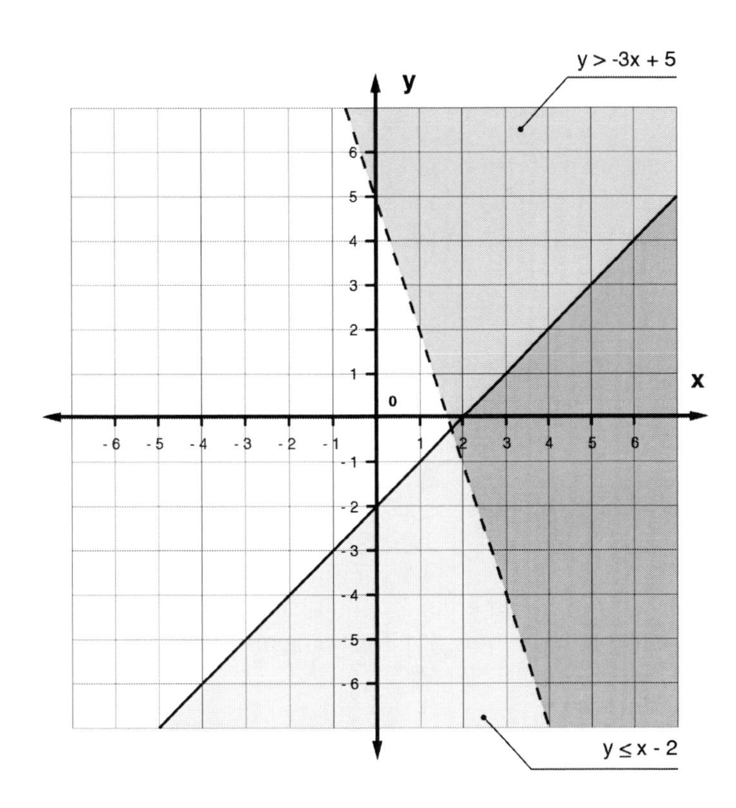

An ordered pair (*x*, *y*) on either side of the line should be chosen to test in the inequality statement. If substituting the values for *x* and *y* results in a true statement ($15(3) + 25(2) > 90$), that ordered pair and all others on that side of the boundary line are part of the solution set. To indicate this, that region of the graph should be shaded. If substituting the ordered pair results in a false statement, the ordered pair and all others on that side are not part of the solution set. Therefore, the other region of the graph contains the solutions and should be shaded.

A question may simply ask whether a given ordered pair is a solution to a given inequality. To determine this, the values should be substituted for the ordered pair into the inequality. If the result is a true statement, the ordered pair is a solution; if the result is a false statement, the ordered pair is not a solution.

Creating Equations and Inequalities to Represent Relationships

An algebraic expression is a statement about unknown quantities expressed in mathematical symbols. The statement *five times a number added to forty* is expressed as $5x + 40$. An equation is a statement in which two expressions (with at least one containing a variable) are equal to one another. The statement *five times a number added to forty is equal to ten* is expressed as:

$$5x + 40 = 10$$

Real world scenarios can also be expressed mathematically. Suppose a job pays its employees $300 per week and $40 for each sale made. The weekly pay is represented by the expression $40x + 300$ where *x* is the number of sales made during the week.

Consider the following scenario: Bob had $20 and Tom had $4. After selling 4 ice cream cones to Bob, Tom has as much money as Bob. The cost of an ice cream cone is an unknown quantity and can be represented by a variable (*x*). The amount of money Bob has after his purchase is four times the cost of an ice cream cone subtracted from his original $\$20 \rightarrow 20 - 4x$. The amount of money Tom has after his sale is four times the cost of an ice cream cone added to his original $\$4 \rightarrow 4x + 4$. After the sale, the amount of money that Bob and Tom have is equal:

$$20 - 4x = 4x + 4$$

Solving either equation for x will yield $4 as the price for an ice cream cone:

$$4x + 4 = 20$$

$$4x = 16$$

$$x = \$4$$

When expressing a verbal or written statement mathematically, it is vital to understand words or phrases that can be represented with symbols. The following are examples:

Symbol	Phrase
+	Added to; increased by; sum of; more than
−	Decreased by; difference between; less than; take away
×	Multiplied by; 3(4,5...) times as large; product of
÷	Divided by; quotient of; half (third, etc.) of
=	Is; the same as; results in; as much as; equal to
x,t,n, etc.	A number; unknown quantity; value of; variable

Use of Formulas

Formulas are mathematical expressions that define the value of one quantity, given the value of one or more different quantities. Formulas look like equations because they contain variables, numbers, operators, and an equal sign. All formulas are equations, but not all equations are formulas. A formula must have more than one variable. For example, $2x + 7 = y$ is an equation and a formula (it relates the unknown quantities *x* and *y*). However, $2x + 7 = 3$ is an equation but not a formula (it only expresses the value of the unknown quantity *x*).

Formulas are typically written with one variable alone (or isolated) on one side of the equal sign. This variable can be thought of as the subject in that the formula is stating the value of the subject in terms of the relationship between the other variables. Consider the distance formula: $distance = rate \times time$ or $d = rt$. The value of the subject variable *d* (distance) is the product of the variable *r* and *t* (rate

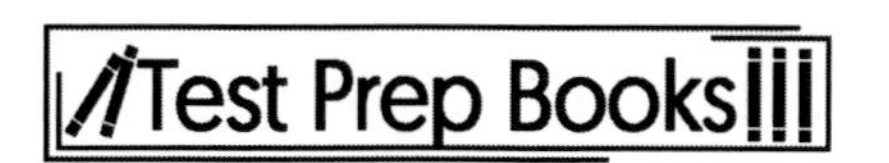

and time). Given the rate and time, the distance traveled can easily be determined by substituting the values into the formula and evaluating.

The formula $P = 2l + 2w$ expresses how to calculate the perimeter of a rectangle (*P*) given its length (*l*) and width (*w*). To find the perimeter of a rectangle with a length of 3ft and a width of 2ft, these values are substituted into the formula for *l* and *w*:

$$P = 2(3ft) + 2(2ft)$$

Following the order of operations, the perimeter is determined to be 10ft. When working with formulas such as these, including units is an important step.

Given a formula expressed in terms of one variable, the formula can be manipulated to express the relationship in terms of any other variable. In other words, the formula can be rearranged to change which variable is the subject. To solve for a variable of interest by manipulating a formula, the equation may be solved as if all other variables were numbers. The same steps for solving are followed, leaving operations in terms of the variables instead of calculating numerical values. For the formula $P = 2l + 2w$, the perimeter is the subject expressed in terms of the length and width. To write a formula to calculate the width of a rectangle, given its length and perimeter, the previous formula relating the three variables is solved for the variable *w*. If *P* and *l* were numerical values, this is a two-step linear equation solved by subtraction and division. To solve the equation $P = 2l + 2w$ for *w*, 2*l* is first subtracted from both sides:

$$P - 2l = 2w$$

Then both sides are divided by 2:

$$\frac{P - 2l}{2} = w$$

Function and Function Notation

A **function** is defined as a relationship between inputs and outputs where there is only one output value for a given input. As an example, the following function is in function notation:

$$f(x) = 3x - 4$$

The $f(x)$ represents the output value for an input of x. If $x = 2$, the equation becomes:

$$f(2) = 3(2) - 4 = 6 - 4 = 2$$

The input of 2 yields an output of 2, forming the ordered pair $(2, 2)$. The following set of ordered pairs corresponds to the given function: $(2, 2), (0, -4), (-2, -10)$. The set of all possible inputs of a function is its domain, and all possible outputs is called the range. By definition, each member of the domain is paired with only one member of the range.

Functions can also be defined recursively. In this form, they are not defined explicitly in terms of variables. Instead, they are defined using previously-evaluated function outputs, starting with either $f(0)$ or $f(1)$. An example of a recursively-defined function is:

$$f(1) = 2, f(n) = 2f(n - 1) + 2n, n > 1$$

The domain of this function is the set of all integers.

Functions can be built out of the context of a situation. For example, the relationship between the money paid for a gym membership and the months that someone has been a member can be described through a function. If the one-time membership fee is $40 and the monthly fee is $30, then the function can be written:

$$f(x) = 30x + 40$$

The x-value represents the number of months the person has been part of the gym, while the output is the total money paid for the membership. The table below shows this relationship. It is a representation of the function because the initial cost is $40 and the cost increases each month by $30.

x (months)	y (money paid to gym)
0	40
1	70
2	100
3	130

Functions can also be built from existing functions. For example, a given function $f(x)$ can be transformed by adding a constant, multiplying by a constant, or changing the input value by a constant. The new function $g(x) = f(x) + k$ represents a vertical shift of the original function. In $f(x) = 3x - 2$, a vertical shift 4 units up would be:

$$g(x) = 3x - 2 + 4 = 3x + 2$$

Multiplying the function times a constant k represents a vertical stretch, based on whether the constant is greater than or less than 1. The function

$$g(x) = kf(x) = 4(3x - 2) = 12x - 8$$

represents a stretch.

Changing the input x by a constant forms the function:

$$g(x) = f(x + k)$$

$$3(x + 4) - 2$$

$$3x + 12 - 2 = 3x + 10$$

and this represents a horizontal shift to the left 4 units. If $(x - 4)$ was plugged into the function, it would represent a vertical shift.

A composition function can also be formed by plugging one function into another. In function notation, this is written:

$$(f\,^\circ g)(x) = f(g(x))$$

For two functions $f(x) = x^2$ and $g(x) = x - 3$, the composition function becomes:

$$f\big(g(x)\big) = (x - 3)^2 = x^2 - 6x + 9$$

The composition of functions can also be used to verify if two functions are inverses of each other. Given the two functions $f(x) = 2x + 5$ and $g(x) = \frac{x-5}{2}$, the composition function can be found $(f°g)(x)$. Solving this equation yields:

$$f(g(x)) = 2\left(\frac{x-5}{2}\right) + 5$$

$$x - 5 + 5 = x$$

It also is true that $g(f(x)) = x$. Since the composition of these two functions gives a simplified answer of x, this verifies that $f(x)$ and $g(x)$ are inverse functions. The domain of $f(g(x))$ is the set of all x-values in the domain of $g(x)$ such that $g(x)$ is in the domain of $f(x)$. Basically, both $f(g(x))$ and $g(x)$ have to be defined.

Functions can also be formed from combinations of existing functions.

Given $f(x)$ and $g(x)$, the following can be built:

$$f + g$$

$$f - g$$

$$fg$$

$$\frac{f}{g}$$

The domains of $f + g, f - g,$ and fg are the intersection of the domains of f and g. The domain of $\frac{f}{g}$ is the same set, excluding those values that make $g(x) = 0$.

For example, if:

$$f(x) = 2x + 3$$

$$g(x) = x + 1$$

then

$$\frac{f}{g} = \frac{2x+3}{x+1}$$

Its domain is all real numbers except -1.

Identifying Zeros of Polynomials

Finding the zeros of polynomial functions is the same process as finding the solutions of polynomial equations. These are the points at which the graph of the function crosses the x-axis. As stated previously, factors can be used to find the zeros of a polynomial function. The degree of the function shows the number of possible zeros. If the highest exponent on the independent variable is 4, then the degree is 4, and the number of possible zeros is 4. If there are complex solutions, the number of roots is less than the degree.

Given the function $y = x^2 + 7x + 6$, y can be set equal to zero, and the polynomial can be factored. The equation turns into $0 = (x + 1)(x + 6)$, where $x = -1$ and $x = -6$ are the zeros. Since this is a quadratic equation, the shape of the graph will be a parabola. Knowing that zeros represent the points where the parabola crosses the x-axis, the maximum or minimum point is the only other piece needed to sketch a rough graph of the function. By looking at the function in standard form, the coefficient of x is positive; therefore, the parabola opens up.

Using the zeros and the minimum, the following rough sketch of the graph can be constructed:

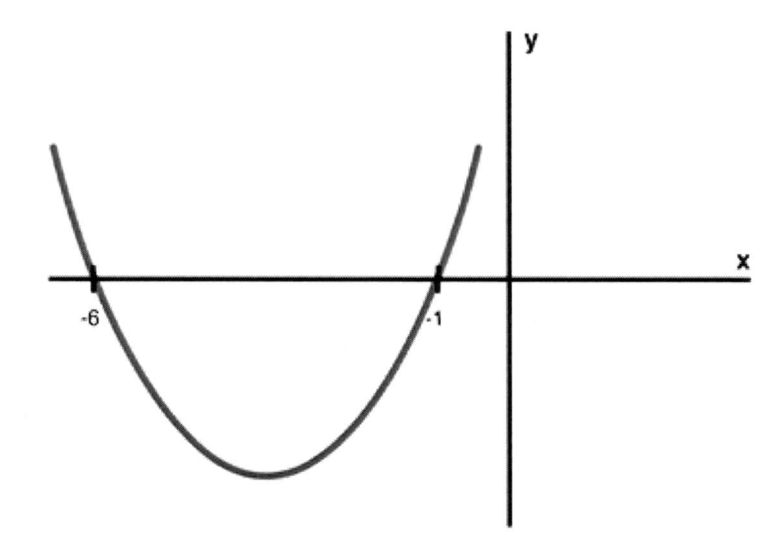

Calculating and Interpreting the Average Rate of Change

Rate of change for any line calculates the steepness of the line over a given interval. Rate of change is also known as the slope or rise/run. The rates of change for nonlinear functions vary depending on the interval being used for the function. The rate of change over one interval may be zero, while the next interval may have a positive rate of change. The equation plotted on the graph below, $y = x^2$, is a quadratic function and non-linear. The average rate of change from points $(0, 0)$ to $(1, 1)$ is 1 because the vertical change is 1 over the horizontal change of 1. For the next interval, $(1, 1)$ to $(2, 4)$, the average rate of change is 3 because the slope is $\frac{3}{1}$.

You can see that here:

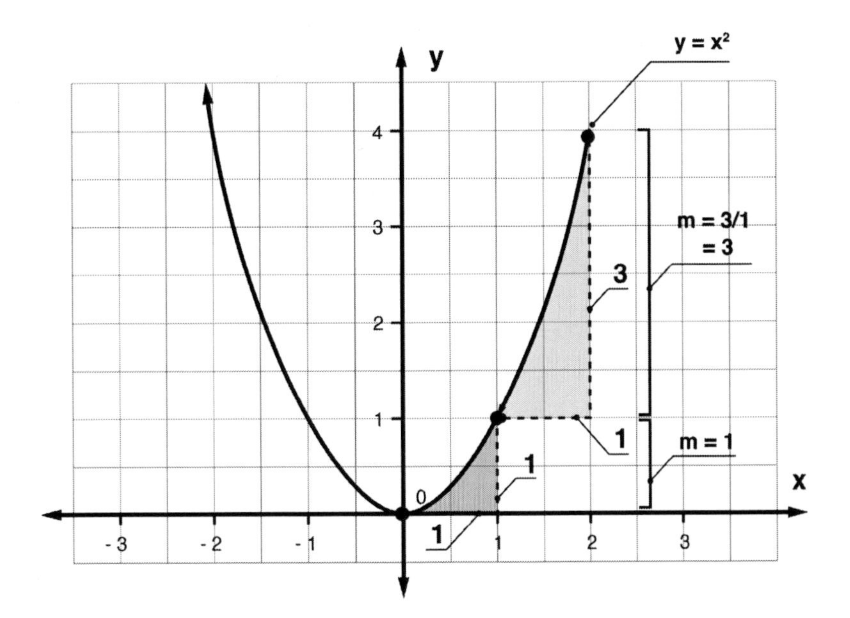

The rate of change for a linear function is constant and can be determined based on a few representations. One method is to place the equation in slope-intercept form: $y = mx + b$. Thus, m is the slope, and b is the y-intercept. In the graph below, the equation is $y = x + 1$, where the slope is 1 and the y-intercept is 1. For every vertical change of 1 unit, there is a horizontal change of 1 unit. The x-intercept is -1, which is the point where the line crosses the x-axis.

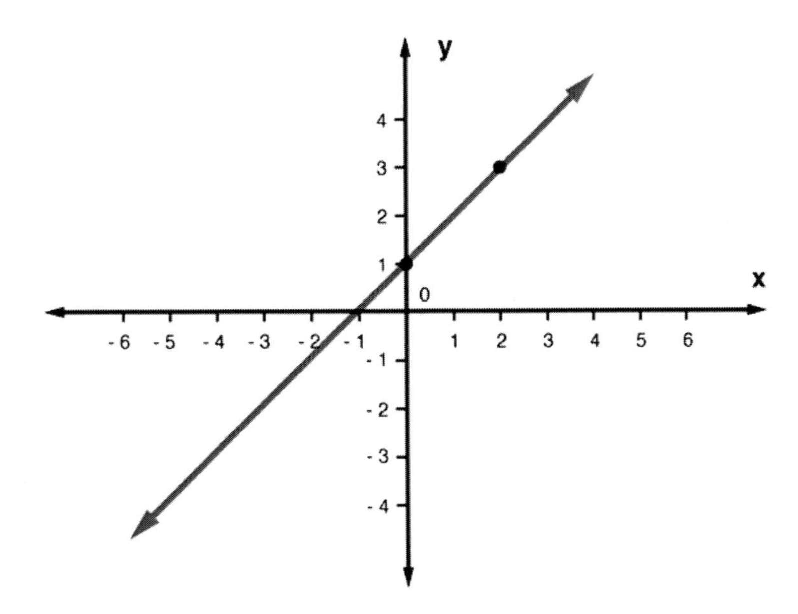

Probability and Statistics

Summarizing and Interpreting Data

Summarizing Data

Most statistics involve collecting a large amount of data, analyzing it, and then making decisions based on previously known information. These decisions also can be measured through additional data collection and then analyzed. Therefore, the cycle can repeat itself over and over. Representing the data visually is a large part of the process, and many plots on the real number line exist that allow this to be done. For example, a **dot plot** uses dots to represent data points above the number line. Also, a **histogram** represents a data set as a collection of rectangles, which illustrate the frequency distribution of the data. Finally, a **box plot** (also known as a box and whisker plot) plots a data set on the number line by segmenting the distribution into four quartiles that are divided equally in half by the median.

Here's an example of a box plot, a histogram, and a dot plot for the same data set:

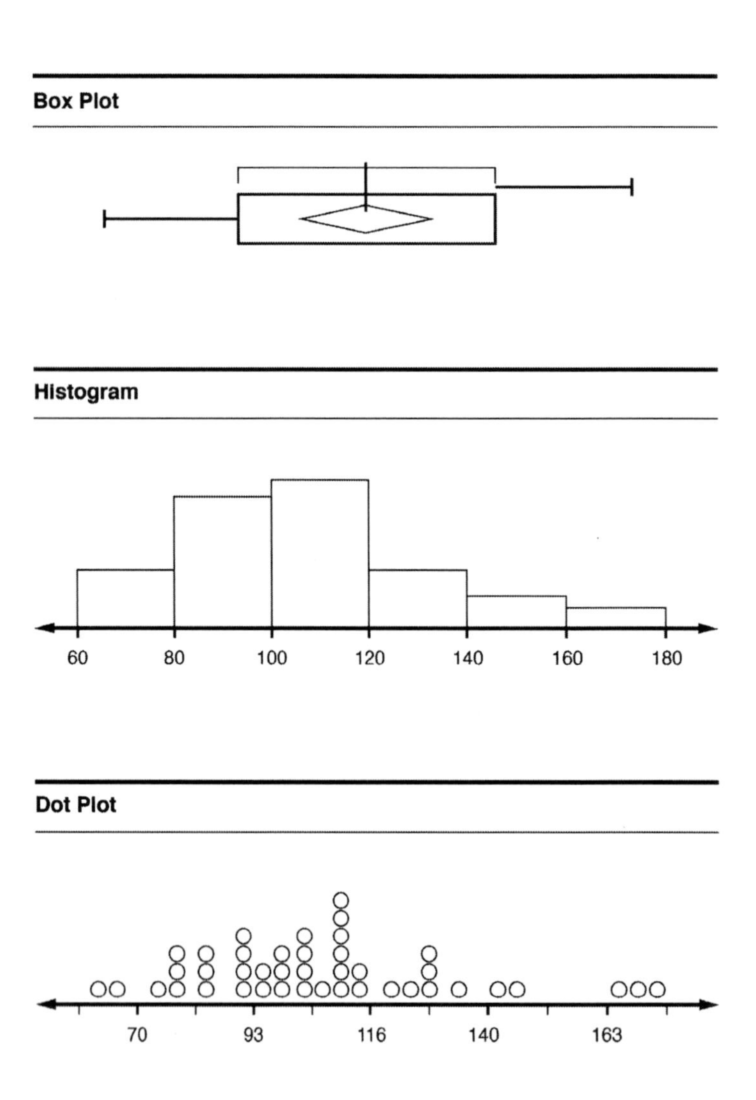

Comparing Data

Comparing data sets within statistics can mean many things. The first way to compare data sets is by looking at the center and spread of each set. The center of a data set can mean two things: median or mean. The **median** is the value that's halfway into each data set, and it splits the data into two intervals. The **mean** is the average value of the data within a set. It's calculated by adding up all of the data in the set and dividing the total by the number of data points. Outliers can significantly impact the mean. Additionally, two completely different data sets can have the same mean. For example, a data set with values ranging from zero to 100 and a data set with values ranging from 44 to 56 can both have means of 50.

The first data set has a much wider range, which is known as the *spread* of the data. This measures how varied the data is within each set. Spread can be defined further as either interquartile range or standard deviation. The **interquartile range (IQR)** is the range of the middle 50 percent of the data set. This range can be seen in the large rectangle on a box plot. The **standard deviation (σ)** quantifies the amount of variation with respect to the mean. A lower standard deviation shows that the data set doesn't differ greatly from the mean. A larger standard deviation shows that the data set is spread out farther from the mean. The formula used for standard deviation depends on whether it's being used for a population or a sample (a subset of a population). The formula for sample standard deviation is:

$$s = \sqrt{\frac{\sum(x - \bar{x})^2}{n - 1}}$$

In this formula, s represents the standard deviation value, x is each value in the data set, $\bar{x}$ is the sample mean, and n is the total number of data points in the set. Note that sample standard deviations use *one less than the total* in the denominator. The population standard deviation formula is similar:

$$\sigma = \sqrt{\frac{\sum(x - \mu)^2}{N}}$$

For population standard deviations, sigma (σ) represents the standard deviation, x represents each value in the data set, mu (μ) is the population mean, and N is the total number of data points for the population.

Interpreting Data

The shape of a data set is another way to compare two or more sets of data. If a data set isn't symmetric around its mean, it's said to be *skewed.* If the tail to the left of the mean is longer, it's said to be skewed to the left. In this case, the mean is less than the median. Conversely, if the tail to the right of the mean is longer, it's said to be skewed to the right and the mean is greater than the median. When classifying a data set according to its shape, its overall skewness is being discussed. If the mean and median are equal, the data set isn't *skewed*; it is *symmetric*, and is considered normally distributed.

An **outlier** is a data point that lies a great distance away from the majority of the data set. It also can be labelled as an extreme value. Technically, an outlier is any value that falls 1.5 times the IQR above the upper quartile or 1.5 times the IQR below the lower quartile. The effect of outliers in the data set is seen visually because they affect the mean. If there's a large difference between the mean and mode, outliers are the cause. The mean shows bias towards the outlying values. However, the median won't be affected as greatly by outliers.

Using Measures of Center to Draw Inferences About Populations

The center of a set of data (statistical values) can be represented by its mean, median, or mode. These are sometimes referred to as measures of central tendency.

Mean

The first property that can be defined for this set of data is the mean. This is the same as the average. To find the mean, add up all the data points, then divide by the total number of data points. For example, suppose that in a class of 10 students, the scores on a test were 50, 60, 65, 65, 75, 80, 85, 85, 90, 100. Therefore, the average test score will be:

$$\frac{50 + 60 + 65 + 65 + 75 + 80 + 85 + 85 + 90 + 100}{10} = 75.5$$

The mean is a useful number if the distribution of data is normal which roughly means that the frequency of different outcomes has a single peak and is roughly equally distributed on both sides of that peak. However, it is less useful in some cases where the data might be split or where there are some outliers. Outliers are data points that are far from the rest of the data. For example, suppose there are 10 executives and 90 employees at a company. The executives make $1000 per hour, and the employees make $10 per hour.

Therefore, the average pay rate will be:

$$\frac{\$1000 \times 10 + \$10 \times 90}{100} = \$109 \text{ per hour}$$

In this case, this average is not very descriptive since it's not close to the actual pay of the executives or the employees.

Median

Another useful measurement is the median. In a data set, the median is the point in the middle. The middle refers to the point where half the data comes before it and half comes after, when the data is recorded in numerical order. For instance, these are the speeds of the fastball of a pitcher during the last inning that he pitched (in order from least to greatest):

90, 92, 93, 93, 95, 96, 97, 97, 97

There are nine total numbers, so the middle or median number is the 5th one, which is 95.

In cases where the number of data points is an even number, then the average of the two middle points is taken. In the previous example of test scores, the two middle points are 75 and 80. Since there is no single point, the average of these two scores needs to be found. The average is:

$$\frac{75 + 80}{2} = 77.5$$

The median is generally a good value to use if there are a few outliers in the data. It prevents those outliers from affecting the "middle" value as much as when using the mean.

Since an outlier is a data point that is far from most of the other data points in a data set, this means an outlier also is any point that is far from the median of the data set. The outliers can have a substantial

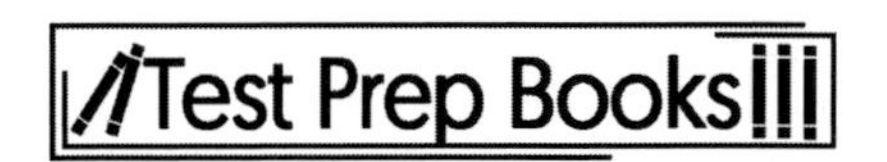

effect on the mean of a data set, but they usually do not change the median or mode, or do not change them by a large quantity. For example, consider the data set (3, 5, 6, 6, 6, 8). This has a median of 6 and a mode of 6, with a mean of $\frac{34}{6} \approx 5.67$. Now, suppose a new data point of 1000 is added so that the data set is now (3, 5, 6, 6, 6, 8, 1000). The median and mode, which are both still 6, remain unchanged. However, the average is now $\frac{1034}{7}$, which is approximately 147.7. In this case, the median and mode will be better descriptions for most of the data points.

Outliers in a given data set are sometimes the result of an error by the experimenter, but oftentimes, they are perfectly valid data points that must be taken into consideration.

Mode

One additional measure to define for X is the mode. This is the data point that appears most frequently. If two or more data points all tie for the most frequent appearance, then each of them is considered a mode. In the case of the test scores, where the numbers were 50, 60, 65, 65, 75, 80, 85, 85, 90, 100, there are two modes: 65 and 85.

Using Statistics to Gain Information About a Population

Statistics involves making decisions and predictions about larger data sets based on smaller data sets. Basically, the information from one part or subset can help predict what happens in the entire data set or population at large. The entire process involves guessing, and the predictions and decisions may not be 100 percent correct all of the time; however, there is some truth to these predictions, and the decisions do have mathematical support. The smaller data set is called a **sample** and the larger data set (in which the decision is being made) is called a **population.** A random sample is used as the sample, which is an unbiased collection of data points that represents the population as well as it can. There are many methods of forming a random sample, and all adhere to the fact that every potential data point has a predetermined probability of being chosen. Statistical inference, based in probability theory, makes calculated assumptions about an entire population based on data from a sample set from that population.

A population is the entire set of people or things of interest. Suppose a study is intended to determine the number of hours of sleep per night for college females in the U.S. The population would consist of every college female in the country. A sample is a subset of the population that may be used for the study. It would not be practical to survey every female college student, so a sample might consist of 100 students per school from 20 different colleges in the country. From the results of the survey, a sample statistic can be calculated. A sample statistic is a numerical characteristic of the sample data, including mean and variance. A sample statistic can be used to estimate a corresponding population parameter. A population parameter is a numerical characteristic of the entire population. Suppose the sample data had a mean (average) of 5.5. This sample statistic can be used as an estimate of the population parameter (average hours of sleep for every college female in the U.S.).

Representing Data

Chart is a broad term that refers to a variety of ways to represent data.

To graph relations, the **Cartesian plane** is used. This means to think of the plane as being given a grid of squares, with one direction being the *x*-axis and the other direction the *y*-axis. Generally, the independent variable is placed along the horizontal axis, and the dependent variable is placed along the vertical axis. Any point on the plane can be specified by saying how far to go along the *x*-axis and how far along the *y*-axis with a pair of numbers (x, y). Specific values for these pairs can be given names such as $C = (-1, 3)$. Negative values mean to move left or down; positive values mean to move right or up. The point where the axes cross one another is called the **origin**. The origin has coordinates $(0, 0)$ and is usually called *O* when given a specific label. An illustration of the Cartesian plane, along with the plotted points $(2, 1)$ and $(-1, -1)$, is below.

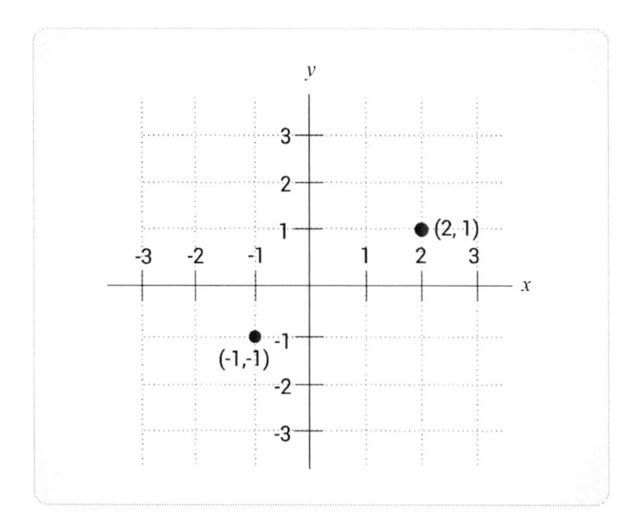

A **line plot** is a diagram that shows quantity of data along a number line. It is a quick way to record data in a structure similar to a bar graph without needing to do the required shading of a bar graph. Here is an example of a line plot:

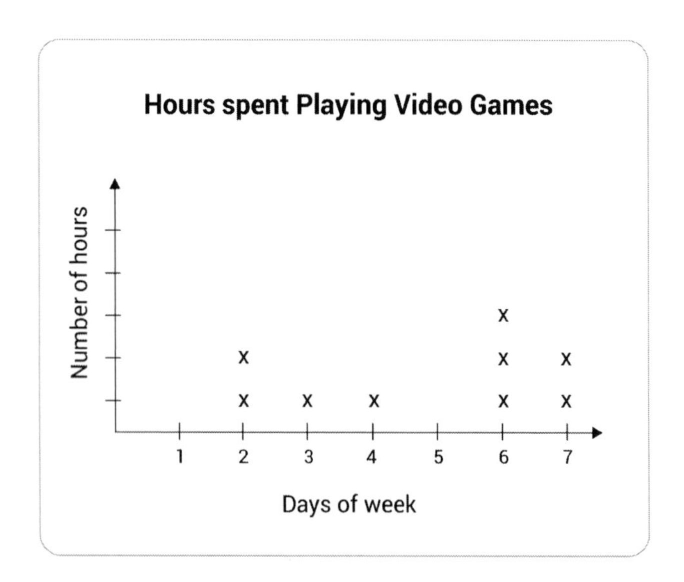

A **tally chart** is a diagram in which tally marks are utilized to represent data. Tally marks are a means of showing a quantity of objects within a specific classification. Here is an example of a tally chart:

Number of days with rain	Number of weeks
0	II
1	~~IIII~~
2	~~IIII~~
3	~~IIII~~
4	~~IIII~~ ~~IIII~~ ~~IIII~~ IIII
5	~~IIII~~ I
6	~~IIII~~ I
7	IIII

Data is often recorded using fractions, such as half a mile, and understanding fractions is critical because of their popular use in real-world applications. Also, it is extremely important to label values with their units when using data. For example, regarding length, the number 2 is meaningless unless it is attached to a unit. Writing 2 cm shows that the number refers to the length of an object.

A **picture graph** is a diagram that shows pictorial representation of data being discussed. The symbols used can represent a certain number of objects. Notice how each fruit symbol in the following graph represents a count of two fruits. One drawback of picture graphs is that they can be less accurate if each symbol represents a large number. For example, if each banana symbol represented ten bananas, and students consumed 22 bananas, it may be challenging to draw and interpret two and one-fifth bananas as a frequency count of 22.

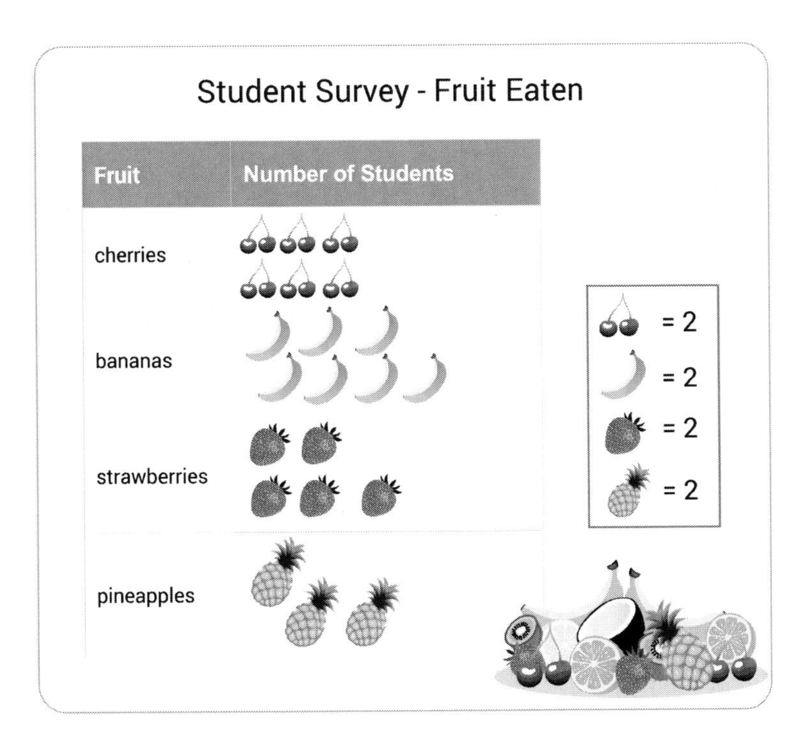

A **circle graph**, also called a pie chart, shows categorical data with each category representing a percentage of the whole data set. To make a circle graph, the percent of the data set for each category must be determined. To do so, the frequency of the category is divided by the total number of data points and converted to a percent. For example, if 80 people were asked what their favorite sport is and 20 responded basketball, basketball makes up 25% of the data:

$$\frac{20}{80} = 0.25 = 25\%$$

Each category in a data set is represented by a *slice* of the circle proportionate to its percentage of the whole.

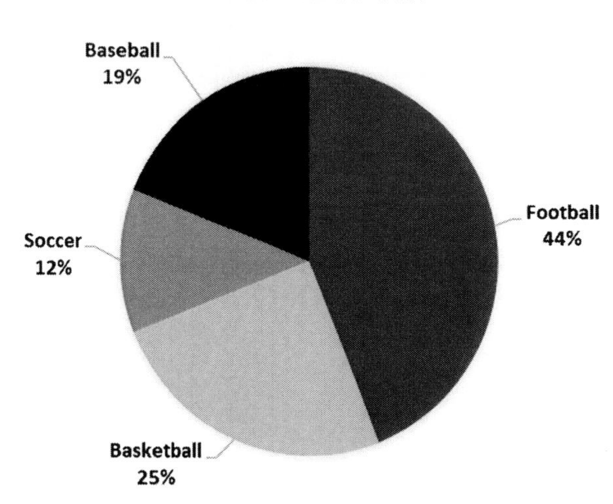

A **scatter plot** displays the relationship between two variables. Values for the independent variable, typically denoted by *x*, are paired with values for the dependent variable, typically denoted by *y*. Each set of corresponding values are written as an ordered pair (*x*, *y*). To construct the graph, a coordinate grid is labeled with the *x*-axis representing the independent variable and the *y*-axis representing the dependent variable. Each ordered pair is graphed.

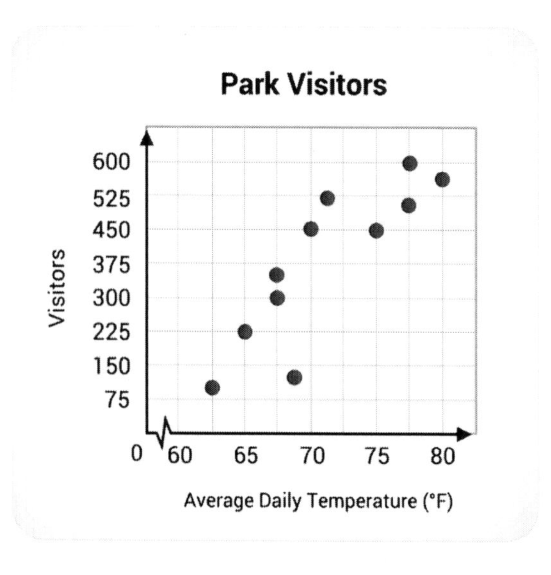

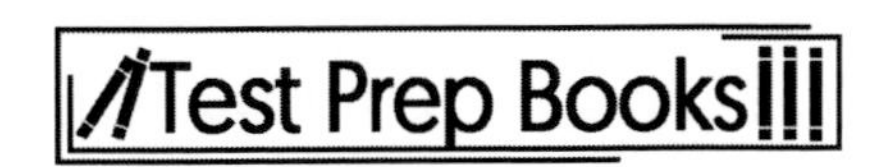

Like a scatter plot, a **line graph** compares two variables that change continuously, typically over time. Paired data values (ordered pair) are plotted on a coordinate grid with the *x*- and *y*-axis representing the two variables. A line is drawn from each point to the next, going from left to right. A double line graph simply displays two sets of data that contain values for the same two variables. The double line graph below displays the profit for given years (two variables) for Company A and Company B (two data sets).

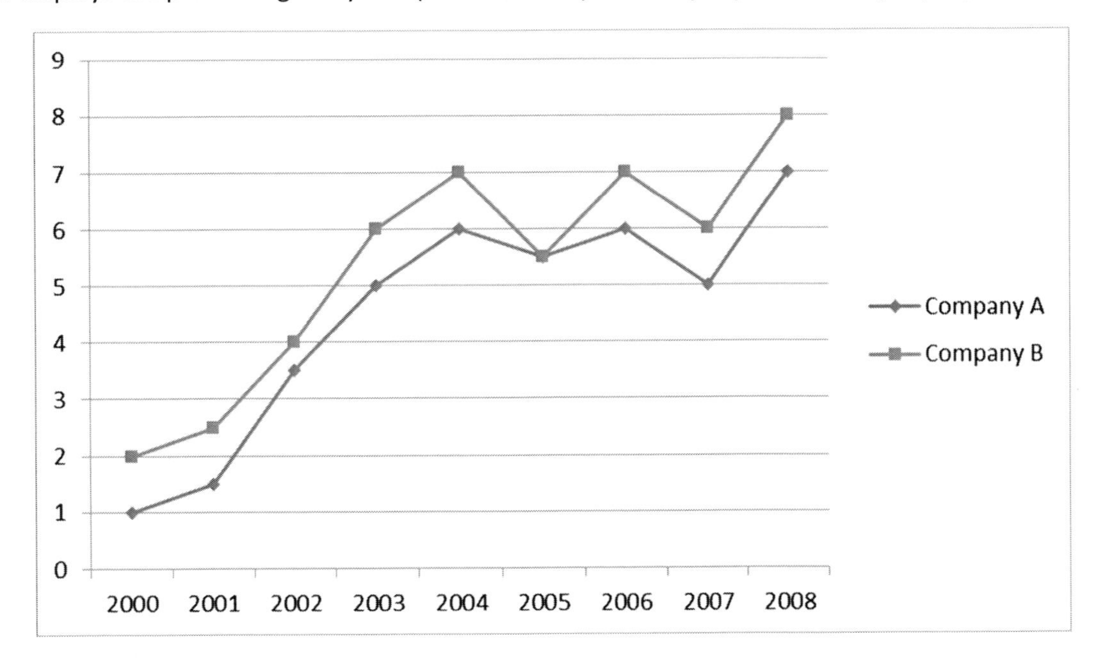

Choosing the appropriate graph to display a data set depends on what type of data is included in the set and what information must be shown.

Scatter plots and line graphs can be used to display data consisting of two variables. Examples include height and weight, or distance and time. A correlation between the variables is determined by examining the points on the graph. Line graphs are used if each value for one variable pairs with a distinct value for the other variable. Line graphs show relationships between variables.

Approximating the Probability of a Chance Event

Probability is a measure of how likely an event is to occur. Probability is written as a fraction between zero and one. If an event has a probability of zero, the event will never occur. If an event has a probability of one, the event will definitely occur. If the probability of an event is closer to zero, the event is unlikely to occur. If the probability of an event is closer to one, the event is more likely to occur. For example, a probability of $\frac{1}{2}$ means that the event is equally as likely to occur as it is not to occur. An example of this is tossing a coin. To calculate the probability of an event, the number of favorable outcomes is divided by the number of total outcomes. For example, suppose you have 2 raffle tickets out of 20 total tickets sold. The probability that you win the raffle is calculated:

$$\frac{number\ of\ favorable\ outcomes}{total\ numberof\ outcomes} = \frac{2}{20} = \frac{1}{10}$$

(always reduce fractions). Therefore, the probability of winning the raffle is $\frac{1}{10}$ or 0.1.

Chance is the measure of how likely an event is to occur, written as a percent. If an event will never occur, the event has a 0% chance. If an event will certainly occur, the event has a 100% chance. If an event will sometimes occur, the event has a chance somewhere between 0% and 100%. To calculate chance, probability is calculated, and the fraction is converted to a percent.

The probability of multiple events occurring can be determined by multiplying the probability of each event. For example, suppose you flip a coin with heads and tails, and roll a six-sided dice numbered one through six. To find the probability that you will flip heads AND roll a two, the probability of each event is determined, and those fractions are multiplied. The probability of flipping heads is:

$$\frac{1}{2}\left(\frac{1\ side\ with\ heads}{2\ sides\ total}\right)$$

and the probability of rolling a two is:

$$\frac{1}{6}\left(\frac{1\ side\ with\ a\ 2}{6\ total\ sides}\right)$$

The probability of flipping heads AND rolling a 2 is:

$$\frac{1}{2} \times \frac{1}{6} = \frac{1}{12}$$

The above scenario with flipping a coin and rolling a dice is an example of independent events. Independent events are circumstances in which the outcome of one event does not affect the outcome of the other event. Conversely, dependent events are ones in which the outcome of one event affects the outcome of the second event. Consider the following scenario: a bag contains 5 black marbles and 5 white marbles. What is the probability of picking 2 black marbles without replacing the marble after the first pick?

The probability of picking a black marble on the first pick is:

$$\frac{5}{10}\left(\frac{5\ black\ marbles}{10\ total\ marbles}\right)$$

Assuming that a black marble was picked, there are now 4 black marbles and 5 white marbles for the second pick. Therefore, the probability of picking a black marble on the second pick is:

$$\frac{4}{9}\left(\frac{4\ black\ marbles}{9\ total\ marbles}\right)$$

To find the probability of picking two black marbles, the probability of each is multiplied:

$$\frac{5}{10} \times \frac{4}{9} = \frac{20}{90} = \frac{2}{9}$$

Probabilities of Single and Compound Events

A **simple event** consists of only one outcome. The most popular simple event is flipping a coin, which results in either heads or tails. A **compound event** results in more than one outcome and consists of more than one simple event. An example of a compound event is flipping a coin while tossing a die. The result is either heads or tails on the coin and a number from one to six on the die. The probability of a simple event is calculated by dividing the number of possible outcomes by the total number of

outcomes. Therefore, the probability of obtaining heads on a coin is $\frac{1}{2}$, and the probability of rolling a 6 on a die is $\frac{1}{6}$. The probability of compound events is calculated using the basic idea of the probability of simple events. If the two events are independent, the probability of one outcome is equal to the product of the probabilities of each simple event. For example, the probability of obtaining heads on a coin and rolling a 6 is equal to:

$$\frac{1}{2} \times \frac{1}{6} = \frac{1}{12}$$

The probability of either A or B occurring is equal to the sum of the probabilities minus the probability that both A and B will occur. Therefore, the probability of obtaining either heads on a coin or rolling a 6 on a die is:

$$\frac{1}{2} + \frac{1}{6} - \frac{1}{12} = \frac{7}{12}$$

The two events aren't mutually exclusive because they can happen at the same time. If two events are mutually exclusive, and the probability of both events occurring at the same time is zero, the probability of event A or B occurring equals the sum of both probabilities. An example of calculating the probability of two mutually exclusive events is determining the probability of pulling a king or a queen from a deck of cards. The two events cannot occur at the same time.

Geometry and Measurements

Using Congruence and Similarity Criteria for Triangles

To prove theorems about triangles, basic definitions involving triangles (e.g., equilateral, isosceles, etc.) need to be known. Proven theorems concerning lines and angles can be applied to prove theorems about triangles. Common theorems to be proved include: the sum of all angles in a triangle equals 180 degrees; the sum of the lengths of two sides of a triangle is greater than the length of the third side; the base angles of an isosceles triangle are congruent; the line segment connecting the midpoint of two sides of a triangle is parallel to the third side and its length is half the length of the third side; and the medians of a triangle all meet at a single point.

Triangle Congruence

There are five theorems to show that triangles are congruent when it's unknown whether each pair of angles and sides are congruent. Each theorem is a shortcut that involves different combinations of sides and angles that must be true for the two triangles to be congruent.

For example, **side-side-side (SSS)** states that if all sides are equal, the triangles are congruent. **Side-angle-side (SAS)** states that if two pairs of sides are equal and the included angles are congruent, then the triangles are congruent. Similarly, **angle-side-angle (ASA)** states that if two pairs of angles are congruent and the included side lengths are equal, the triangles are similar. **Angle-angle-side (AAS)** states that two triangles are congruent if they have two pairs of congruent angles and a pair of corresponding equal side lengths that aren't included. Finally, **hypotenuse-leg (HL)** states that if two right triangles have equal hypotenuses and an equal pair of shorter sides, then the triangles are congruent.

An important item to note is that angle-angle-angle *(AAA)* is not enough information to have congruence. It's important to understand why these rules work by using rigid motions to show congruence between the triangles with the given properties. For example, three reflections are needed to show why *SAS* follows from the definition of congruence.

Similarity for Two Triangles

If two angles of one triangle are congruent with two angles of a second triangle, the triangles are similar. This is because, within any triangle, the sum of the angle measurements is 180 degrees. Therefore, if two are congruent, the third angle must also be congruent because their measurements are equal. Three congruent pairs of angles mean that the triangles are similar.

Proving Congruence and Similarity

The criteria needed to prove triangles are congruent involves both angle and side congruence. Both pairs of related angles and sides need to be of the same measurement to use congruence in a proof. The criteria to prove similarity in triangles involves proportionality of side lengths. Angles must be congruent in similar triangles; however, corresponding side lengths only need to be a constant multiple of each other. Once similarity is established, it can be used in proofs as well. Relationships in geometric figures other than triangles can be proven using triangle congruence and similarity. If a similar or congruent triangle can be found within another type of geometric figure, their criteria can be used to prove a relationship about a given formula. For example, a rectangle can be broken up into two congruent triangles.

Properties of Polygons and Circles

A **polygon** is a closed two-dimensional figure consisting of three or more sides. Polygons can be either convex or concave. A polygon that has interior angles all measuring less than 180° is convex. A concave polygon has one or more interior angles measuring greater than 180°. Examples are shown below.

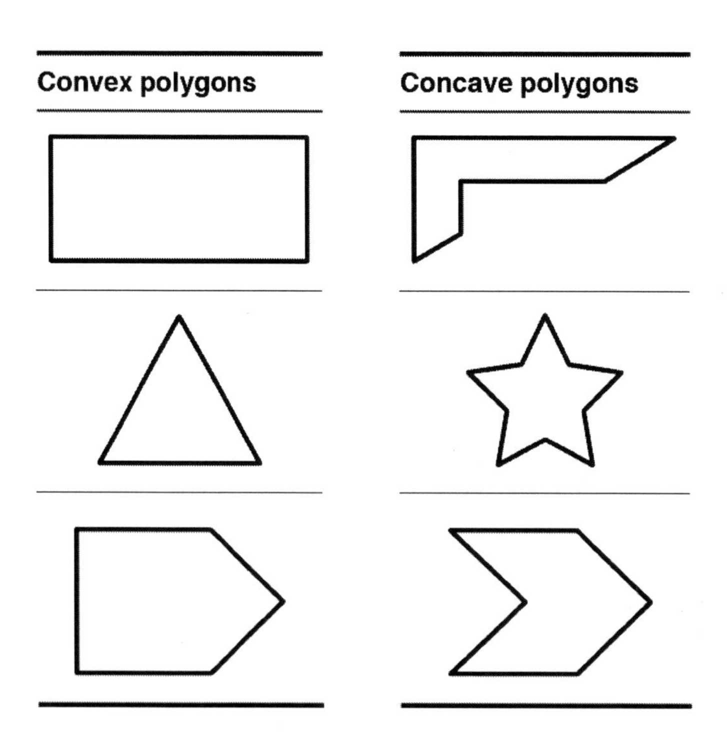

Polygons can be classified by the number of sides (also equal to the number of angles) they have. The following are the names of polygons with a given number of sides or angles:

# of Sides	Name of Polygon
Triangle	3
Quadrilateral	4
Pentagon	5
Hexagon	6
Septagon (or heptagon)	7
Octagon	8
Nonagon	9
Decagon	10

Equiangular polygons are polygons in which the measure of every interior angle is the same. The sides of equilateral polygons are always the same length. If a polygon is both equiangular and equilateral, the polygon is defined as a regular polygon. Examples are shown below.

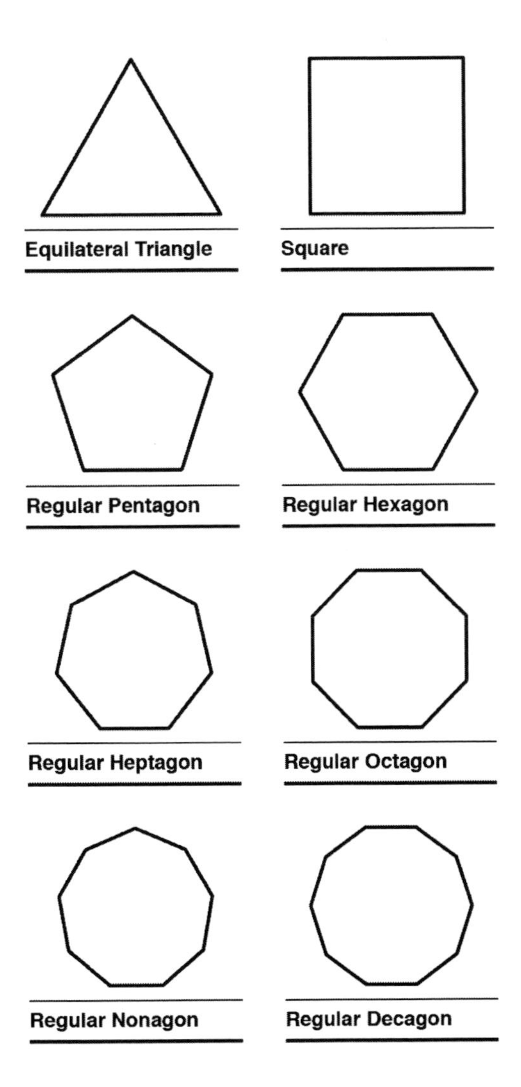

Triangles can be further classified by their sides and angles. A triangle with its largest angle measuring 90° is a right triangle.

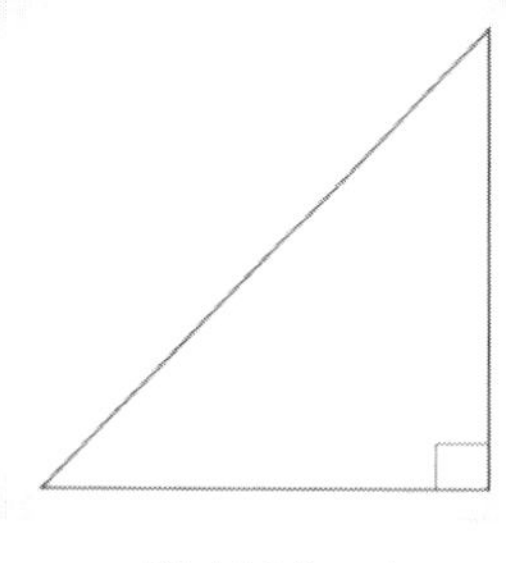

Right triangle

A triangle with the largest angle less than 90° is an acute triangle. A triangle with the largest angle greater than 90° is an obtuse triangle.

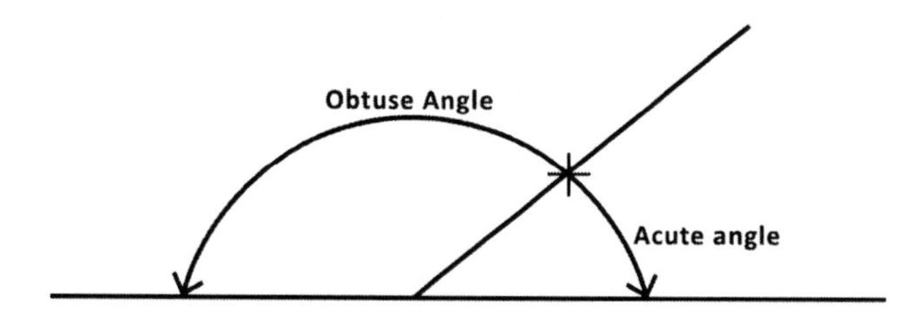

A triangle consisting of two equal sides and two equal angles is an isosceles triangle. A triangle with three equal sides and three equal angles is an equilateral triangle. A triangle with no equal sides or angles is a scalene triangle.

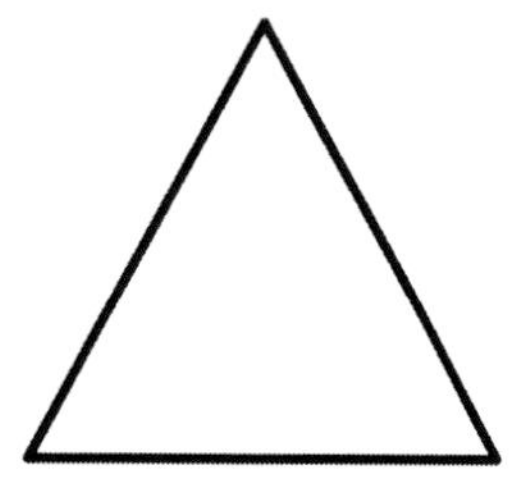

Equilateral Triangle

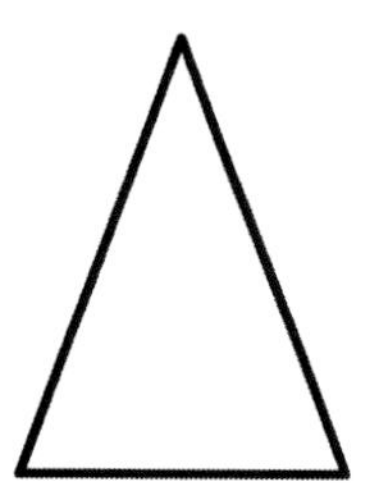

Isosceles Triangle

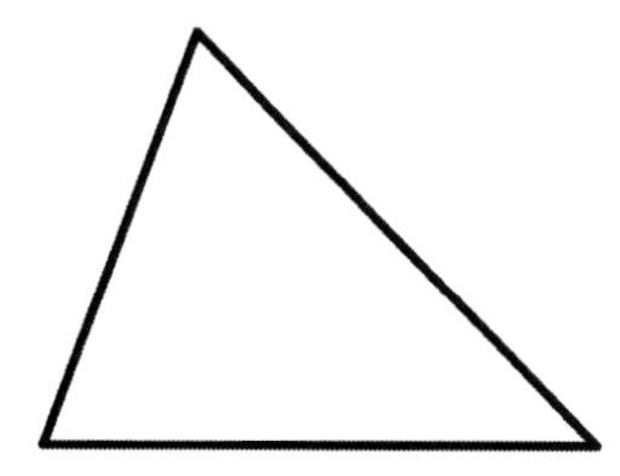

Scalene Triangle

Quadrilaterals can be further classified according to their sides and angles. A quadrilateral with exactly one pair of parallel sides is called a **trapezoid**. A quadrilateral that shows both pairs of opposite sides parallel is a **parallelogram**. Parallelograms include rhombuses, rectangles, and squares. A rhombus has four equal sides. A rectangle has four equal angles (90° each). A square has four 90° angles and four equal sides. Therefore, a square is both a rhombus and a rectangle.

Use this chart for a better understanding of quadrilaterals:

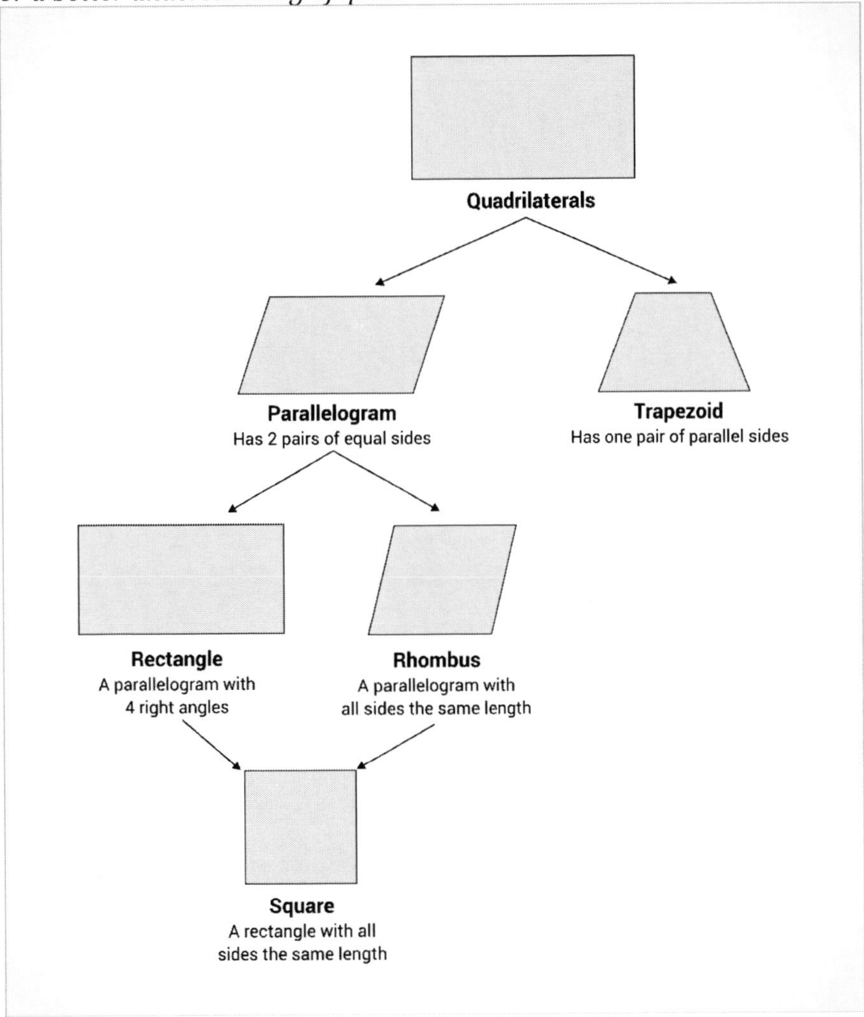

There are many key facts related to geometry that are applicable. The sum of the measures of the angles of a triangle are 180°, and for a quadrilateral, the sum is 360°. Rectangles and squares each have four right angles. A right angle has a measure of 90°.

A parallelogram has six important properties:

- Opposite sides are congruent.
- Opposite angles are congruent.
- Within a parallelogram, consecutive angles are supplementary, so their measurements total 180 degrees.
- If one angle is a right angle, all of them have to be right angles.
- The diagonals of the angles bisect each other.
- These diagonals form two congruent triangles.

Perimeter and Area

The **perimeter** is the distance around a figure or the sum of all sides of a polygon.

The formula for the perimeter of a square is four times the length of a side. For example, the following square has side lengths of 5 meters:

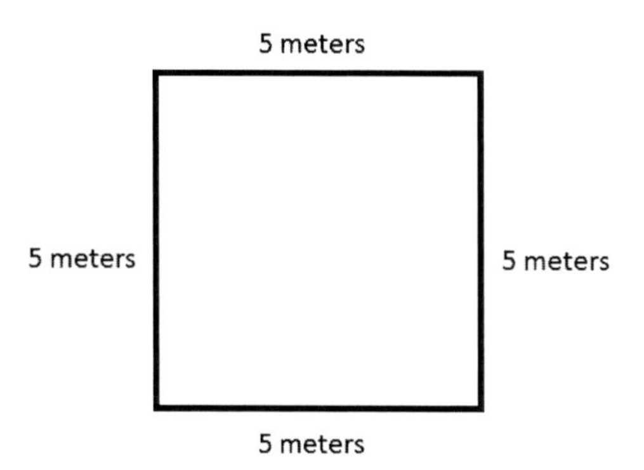

The perimeter is 20 meters because 4 times 5 is 20.

The formula for a perimeter of a rectangle is the sum of twice the length and twice the width. For example, if the length of a rectangle is 10 inches and the width 8 inches, then the perimeter is 36 inches because:

$$P = 2l + 2w = 2(10) + 2(8) = 20 + 16 = 36 \text{ inches}$$

The **circumference** of a circle is found by calculating the perimeter of the shape. The ratio of the circumference to the diameter of the circle is π, so the formula for circumference is:

$$C = \pi d = 2\pi r$$

The **area** is the amount of space inside of a figure, and there are formulas associated with area.

The area of a triangle is the product of ½ the base and height. For example, if the base of the triangle is 2 feet and the height is 4 feet, then the area is 4 square feet. The following equation shows the formula used to calculate the area of the triangle:

$$A = \frac{1}{2}bh = \frac{1}{2}(2)(4) = 4 \text{ square feet}$$

The area of a square is the length of a side squared. For example, if a side of a square is 7 centimeters, then the area is 49 square centimeters. The formula for this example is $A = s^2 = 7^2 = 49$ square centimeters. An example is if the rectangle has a length of 6 inches and a width of 7 inches, then the area is 42 square inches:

$$A = lw = 6(7) = 42 \text{ square inches}$$

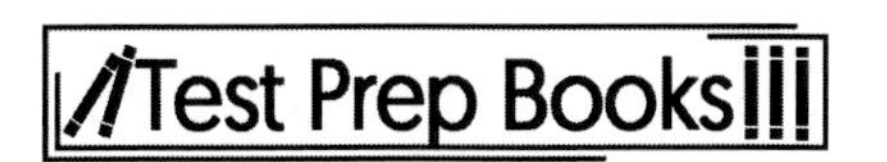

The area of a trapezoid is ½ the height times the sum of the bases. For example, if the length of the bases are 2.5 and 3 feet and the height 3.5 feet, then the area is 9.625 square feet. The following formula shows how the area is calculated:

$$A = \frac{1}{2}h(b_1 + b_2)$$

$$\frac{1}{2}(3.5)(2.5 + 3)$$

$$\frac{1}{2}(3.5)(5.5)$$

$$9.625 \text{ square feet}$$

The perimeter of a figure is measured in single units, while the area is measured in square units.

Surface Area and Volume Formulas

Surface area and volume are two- and three-dimensional measurements. **Surface area** measures the total surface space of an object, like the six sides of a cube. Questions about surface area will ask how much of something is needed to cover a three-dimensional object, like wrapping a present. **Volume** is the measurement of how much space an object occupies, like how much space is in the cube. Volume questions will ask how much of something is needed to completely fill the object. The most common surface area and volume questions deal with spheres, cubes, and rectangular prisms.

The formula for a cube's surface area is $SA = 6 \times s^2$, where s is the length of a side. A cube has 6 equal sides, so the formula expresses the area of all the sides. Volume is simply measured by taking the cube of the length, so the formula is $V = s^3$.

The surface area formula for a rectangular prism or a general box is:

$$SA = 2(lw + lh + wh)$$

l is the length, h is the height, and w is the width. The volume formula is $V = l \times w \times h$, which is the cube's volume formula adjusted for the unequal lengths of a box's sides.

The formula for a sphere's surface area is $SA = 4\pi r^2$, where r is the sphere's radius. The surface area formula is the area for a circle multiplied by four. To measure volume, the formula is:

$$V = \frac{4}{3}\pi r^3$$

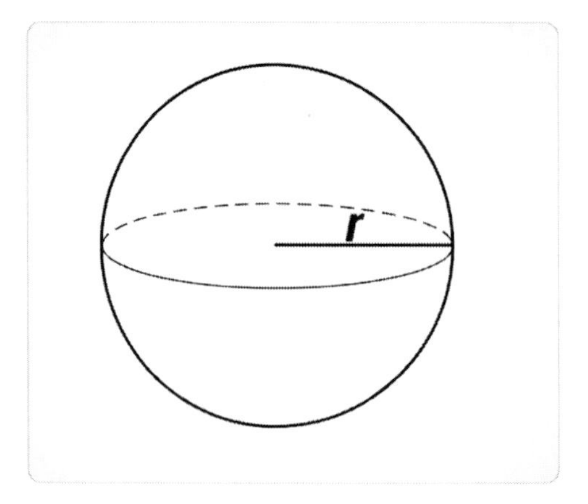

A rectangular pyramid is a figure with a rectangular base and four triangular sides that meet at a single vertex. If the rectangle has sides of lengths x and y, then the volume will be given by:

$$V = \frac{1}{3}xyh$$

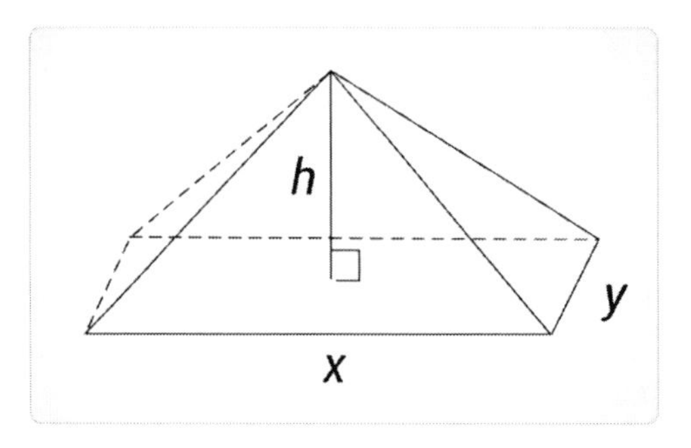

To find the surface area, the dimensions of each triangle must be known. However, these dimensions can differ depending on the problem in question.

Therefore, there is no general formula for calculating total surface area.

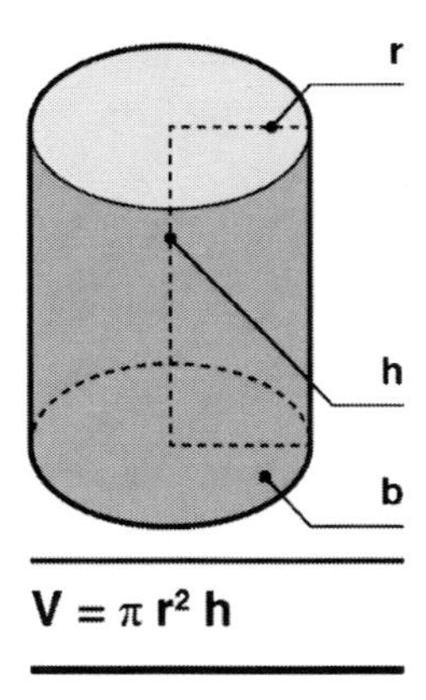

The formula to find the volume of a cylinder is $\pi r^2 h$. This formula contains the formula for the area of a circle (πr^2) because the base of a cylinder is a circle. To calculate the volume of a cylinder, the slices of circles needed to build the entire height of the cylinder are added together. For example, if the radius is 5 feet and the height of the cylinder is 10 feet, the cylinder's volume is calculated by using the following equation: $\pi 5^2 \times 10$. Substituting 3.14 for π, the volume is 785 ft^3.

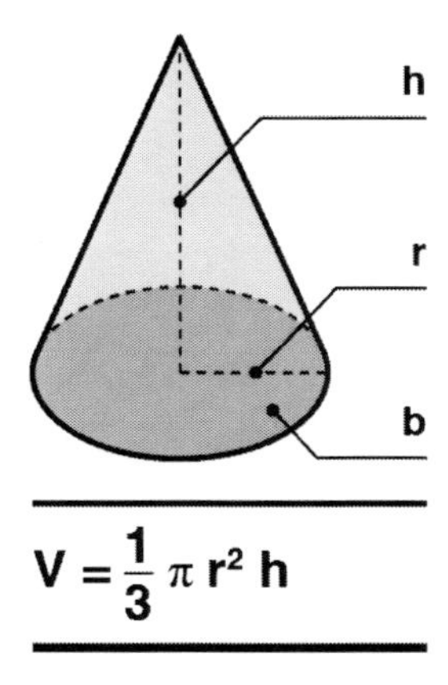

The formula used to calculate the volume of a cone is $\frac{1}{3}\pi r^2 h$. Essentially, the area of the base of the cone is multiplied by the cone's height. In a real-life example where the radius of a cone is 2 meters and the height of a cone is 5 meters, the volume of the cone is calculated by utilizing the formula:

$$\frac{1}{3}\pi 2^2 \times 5$$

After substituting 3.14 for π, the volume is 20.9 m^3.

Angles and Measurement

Supplementary angles add up to 180 degrees. **Vertical angles** are two nonadjacent angles formed by two intersecting lines. For example, in the following picture, angles 4 and 2 are vertical angles and so are angles 1 and 3:

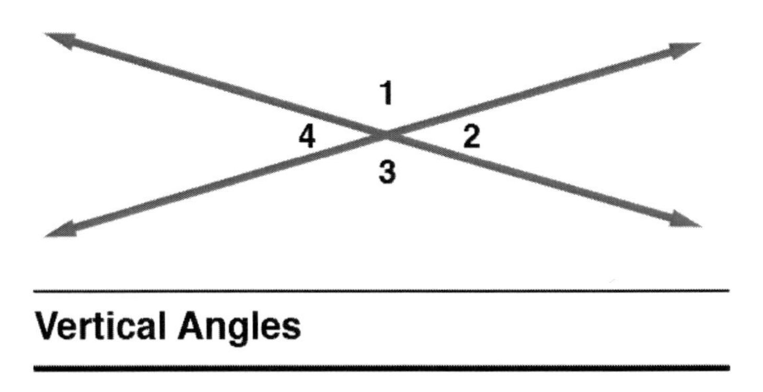

Vertical Angles

Corresponding angles are two angles in the same position whenever a straight line (known as a transversal) crosses two others. If the two lines are parallel, the corresponding angles are equal. In the following diagram, angles 1 and 3 are corresponding angles but aren't equal to each other:

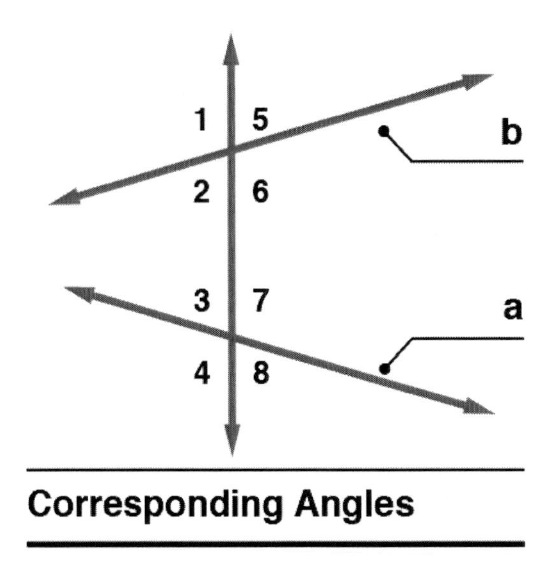

Corresponding Angles

Alternate interior angles are also a pair of angles formed when two lines are crossed by a transversal. They are opposite angles that exist inside of the two lines. In the corresponding angles diagram above, angles 2 and 7 are alternate interior angles, as well as angles 6 and 3. **Alternate exterior angles** are opposite angles formed by a transversal but, in contrast to interior angles, exterior angles exist outside the two original lines. Therefore, angles 1 and 8 are alternate exterior angles and so are angles 5 and 4. Finally, **consecutive interior angles** are pairs of angles formed by a transversal. These angles are located on the same side of the transversal and inside the two original lines. Therefore, angles 2 and 3 are a pair of consecutive interior angles, and so are angles 6 and 7. These definitions are instrumental in solving many problems that involve determining relationships between angles.

The Pythagorean Theorem

The **Pythagorean theorem** is an important relationship between the three sides of a right triangle. It states that the square of the side opposite the right triangle, known as the **hypotenuse** (denoted as c^2), is equal to the sum of the squares of the other two sides ($a^2 + b^2$). Thus,

$$a^2 + b^2 = c^2$$

The theorem can be seen in the following image:

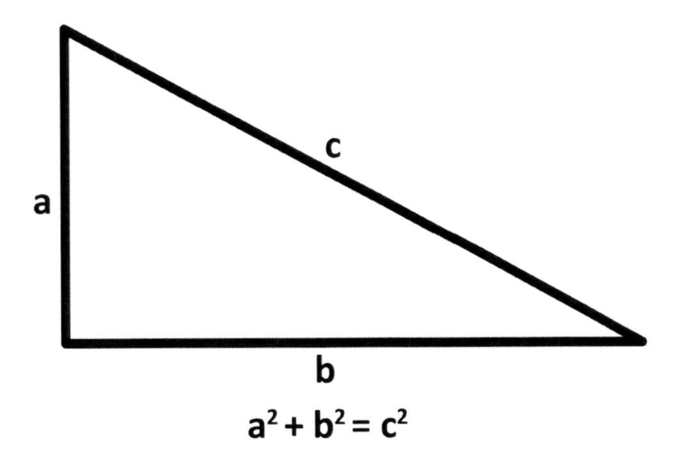

$a^2 + b^2 = c^2$

Both the trigonometric functions and the Pythagorean theorem can be used in problems that involve finding either a missing side or a missing angle of a right triangle. To do so, one must look to see what sides and angles are given and select the correct relationship that will help find the missing value. These relationships can also be used to solve application problems involving right triangles. Often, it's helpful to draw a figure to represent the problem to see what's missing.

As an example of the theorem, suppose that Shirley has a rectangular field that is 5 feet wide and 12 feet long, and she wants to split it in half using a fence that goes from one corner to the opposite corner. How long will this fence need to be? To figure this out, note that this makes the field into two right triangles, whose hypotenuse will be the fence dividing it in half. Therefore, the fence length is given by:

$$\sqrt{5^2 + 12^2} = \sqrt{169} = 13 \text{ feet long}$$

Transformations in the Plane

Transformations in the Plane

A **transformation** occurs when a shape is altered in the plane where it exists. There are three major types of transformation: translations, reflections, and rotations. A **translation** consists of shifting a shape in one direction. A **reflection** results when a shape is transformed over a line to its mirror image. Finally, a **rotation** occurs when a shape moves in a circular motion around a specified point. The object can be turned clockwise or counterclockwise and, if rotated 360 degrees, returns to its original location.

Distance and Angle Measure

The three major types of transformations preserve distance and angle measurement. The shapes stay the same, but they are moved to another place in the plane. Therefore, the distance between any two points on the shape doesn't change. Also, any original angle measure between two line segments doesn't change. However, there are transformations that don't preserve distance and angle measurements, including those that don't preserve the original shape. For example, transformations that involve stretching and shrinking shapes don't preserve distance and angle measures. In these cases, the input variables are multiplied by either a number greater than one (stretch) or less than one (shrink).

Rigid Motion

A **rigid motion** is a transformation that preserves distance and length. Every line segment in the resulting image is congruent to the corresponding line segment in the pre-image. Congruence between two figures means a series of transformations (or a rigid motion) can be defined that maps one of the figures onto the other. Basically, two figures are congruent if they have the same shape and size.

Dilation

A shape is dilated, or a **dilation** occurs, when each side of the original image is multiplied by a given scale factor. If the scale factor is less than 1 and greater than 0, the dilation contracts the shape and the resulting shape is smaller. If the scale factor equals 1, the resulting shape is the same size and the dilation is a rigid motion. Finally, if the scale factor is greater than 1, the resulting shape is larger and the dilation expands the shape. The center of dilation is the point where the distance from it to any point on the new shape equals the scale factor times the distance from the center to the corresponding point in the pre-image. Dilation isn't an isometric transformation because distance isn't preserved. However, angle measure, parallel lines, and points on a line all remain unchanged. This following figure is an example of translation, rotation, dilation, and reflection:

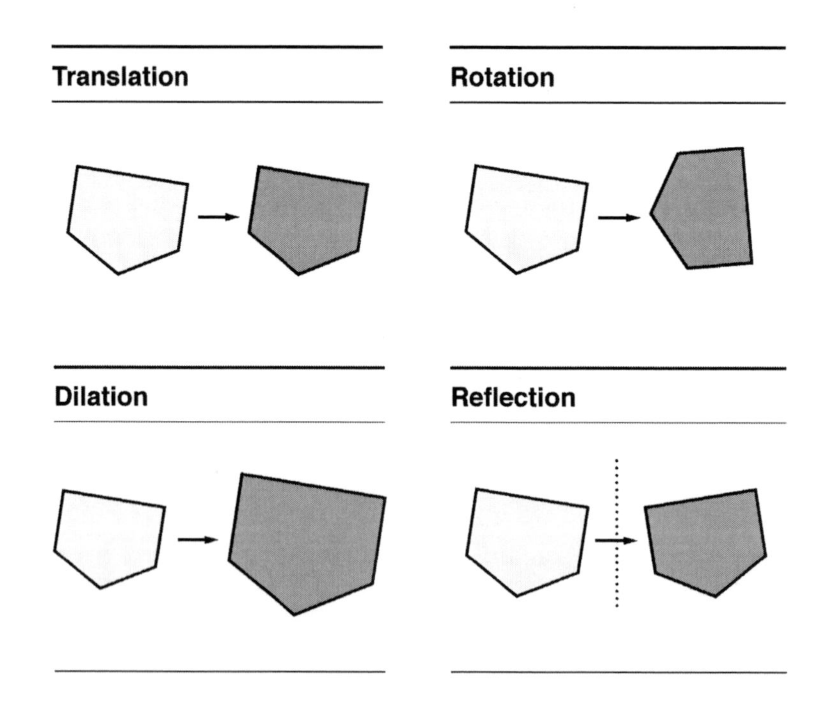

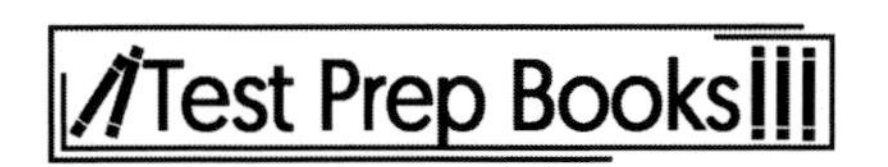

Determining Congruence

Two figures are congruent if there is a rigid motion that can map one figure onto the other. Therefore, all pairs of sides and angles within the image and pre-image must be congruent. For example, in triangles, each pair of the three sides and three angles must be congruent. Similarly, in two four-sided figures, each pair of the four sides and four angles must be congruent.

Properties of Lines

In geometry, a line connects two points, has no thickness, and extends indefinitely in both directions beyond each point. If the length is finite, it's known as a line segment and has two endpoints. A ray is the straight portion of a line that has one endpoint and extends indefinitely in the other direction. An angle is formed when two rays begin at the same endpoint and extend indefinitely. The endpoint of an angle is called a vertex. Adjacent angles are two side-by-side angles formed from the same ray that have the same endpoint. Angles are measured in degrees or radians, which is a measure of rotation. A full rotation equals 360 degrees or 2π radians, which represents a circle. Half a rotation equals 180 degrees or π radians and represents a half-circle. Subsequently, 90 degrees ($\frac{\pi}{2}$ radians) represents a quarter of a circle, which is known as a right angle. Any angle less than 90 degrees is an acute angle, and any angle greater than 90 degrees is an obtuse angle. Angle measurement is additive. When an angle is broken into two non-overlapping angles, the total measure of the larger angle equals the sum of the two smaller angles. Lines are coplanar if they're located in the same plane. Two lines are parallel if they are coplanar, extend in the same direction, and never cross. If lines do cross, they're labeled as intersecting lines because they "intersect" at one point. If they intersect at more than one point, they're the same line. Perpendicular lines are coplanar lines that form a right angle at their point of intersection.

Two lines are parallel if they have the same slope and a different intercept. Two lines are perpendicular if the product of their slope equals -1. Parallel lines never intersect unless they are the same line, and perpendicular lines intersect at a right angle. If two lines aren't parallel, they must intersect at one point. Determining equations of lines based on properties of parallel and perpendicular lines appears in word problems. To find an equation of a line, both the slope and a point the line goes through are necessary. Therefore, if an equation of a line is needed that's parallel to a given line and runs through a specified point, the slope of the given line and the point are plugged into the point-slope form of an equation of a line. Secondly, if an equation of a line is needed that's perpendicular to a given line running through a specified point, the negative reciprocal of the slope of the given line and the point are plugged into the point-slope form. Also, if the point of intersection of two lines is known, that point will be used to solve the set of equations. Therefore, to solve a system of equations, the point of intersection must be found. If a set of two equations with two unknown variables has no solution, the lines are parallel.

SHSAT Practice Test #1

Editing/Revising

Editing/Revising Part A

1. Read this sentence.

   Protestors filled the streets of the city. Because they were dissatisfied with the government's leadership.

How should this sentence be revised?

a. Protestors filled the streets of the city, because they were dissatisfied with the government's leadership.
b. Protesters, filled the streets of the city, because they were dissatisfied with the government's leadership.
c. Because they were dissatisfied with the government's leadership protestors filled the streets of the city.
d. Protestors filled the streets of the city because they were dissatisfied with the government's leadership.

2. Read this sentence.

   She's looking for a suitcase that can fit all of her clothes, shoes, accessory, and makeup.

How should the sentence be revised?

e. Change ***shoes*** to **shoe**.
f. Change the commas to semicolons.
g. Change ***makeup*** to **makeups**.
h. Change ***accessory*** to **accessories**.

3. Read this paragraph.

   (1) Early in my career, a master's teacher shared this thought with me: "Education is the last bastion of civility." (2) While I did not completely understand the scope of those words at the time, I have since come to realize the depth, breadth, truth, and significance of what he said. (3) Education provides society with a vehicle for raising it's children to be civil, decent, human beings with something valuable to contribute to the world. (4) It is really what makes us human and what distinguishes us as civilized creatures.

How should the paragraph be revised?

a. Sentence 1: Move the period to outside the quotation marks.
b. Sentence 2: Remove the comma after ***time***.
c. Sentence 3: Remove the apostrophe from ***it's***.
d. Sentence 4: Change ***distinguishes*** to **distinguished**.

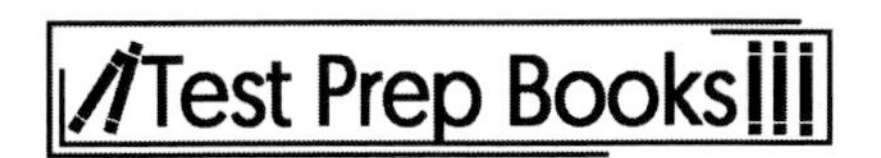

4. Read this paragraph.

> (1) George Washington Carver was an innovator, always thinking of new and better ways to do things and is most famous for his over three hundred uses for the peanut. (2) Toward the end of his career, Carver returns to his first love of art. (3) When Carver died, he left his money to help fund ongoing agricultural research. (4) Today, people still visit and study at the George Washington Carver Foundation at the Tuskegee Institute.

Which sentence contains an error in construction and should be revised?

e. Sentence 1
f. Sentence 2
g. Sentence 3
h. Sentence 4

Editing/Revising Part B

Questions 5–19 are based on the following passage:

> The knowledge of an aircraft engineer is acquired through years of education, and special (1) <u>licenses are required</u>. Ideally, an individual will begin his or her preparation for the profession in high school (2) <u>by taking chemistry physics trigonometry and calculus.</u> Such curricula will aid in (3) <u>one's pursuit of a bachelor's degree</u> in aircraft engineering, which requires several physical and life sciences, mathematics, and design courses.
>
> (4) <u>Some of universities provide internship or apprentice opportunities</u> for the students enrolled in aircraft engineer programs. A bachelor's in aircraft engineering is commonly accompanied by a master's degree in advanced engineering or business administration. Such advanced degrees enable an individual to position himself or herself for executive, faculty, and/or research opportunities. (5) <u>These advanced offices oftentimes require a Professional Engineering (PE) license which can be obtained through additional college courses, professional experience, and acceptable scores on the Fundamentals of Engineering (FE) and Professional Engineering (PE) standardized assessments.</u>
>
> (6) <u>Once the job begins, this lines of work</u> requires critical thinking, business skills, problem solving, and creativity. This level of (7) <u>expertise</u> (8) <u>allows</u> aircraft engineers to (9) <u>apply mathematical equation and scientific processes</u> to aeronautical and aerospace issues or inventions. (10) <u>For example,</u> aircraft engineers may test, design, and construct flying vessels such as airplanes, space shuttles, and missile weapons. As a result, aircraft engineers are compensated with generous salaries. In fact, in May 2014, the lowest 10 percent of all American aircraft engineers earned less than $60,110 while the highest paid ten-percent of all American aircraft engineers earned $155,240. (11) <u>In May 2015, the United States Bureau of Labor Statistics (BLS) reported that the median annual salary of aircraft engineers was $107,830.</u> (12) <u>Conversely</u>, (13) <u>employment opportunities for aircraft engineers are projected to decrease by 2 percent by 2024.</u> This decrease may be the result of a decline in the manufacturing industry. (14) <u>Nevertheless aircraft engineers who know how to utilize</u> modeling and simulation programs, fluid dynamic software, and robotic engineering tools (15) <u>is projected to remain</u> the most employable.

5. Which of the following would be the best choice for this sentence (reproduced below)?

The knowledge of an aircraft engineer is acquired through years of education, and special (1) licenses are required.

a. NO CHANGE
b. licenses will be required
c. licenses may be required
d. licenses should be required

6. Which of the following would be the best choice for this sentence (reproduced below)?

Ideally, an individual will begin his or her preparation for the profession in high school (2) by taking chemistry physics trigonometry and calculus.

e. NO CHANGE
f. by taking chemistry; physics; trigonometry; and calculus.
g. by taking chemistry, physics, trigonometry, and calculus.
h. by taking chemistry, physics, trigonometry, calculus.

7. Which of the following would be the best choice for this sentence (reproduced below)?

Such curricula will aid in (3) one's pursuit of a bachelor's degree in aircraft engineering, which requires several physical and life sciences, mathematics, and design courses.

a. NO CHANGE
b. ones pursuit of a bachelors degree
c. one's pursuit of a bachelors degree
d. ones pursuit of a bachelor's degree

8. Which of the following would be the best choice for this sentence (reproduced below)?

(4) Some of universities provide internship or apprentice opportunities for the students enrolled in aircraft engineer programs.

e. NO CHANGE
f. Some of universities provided internship or apprentice opportunities
g. Some of universities provide internship or apprenticeship opportunities
h. Some universities provide internship or apprenticeship opportunities

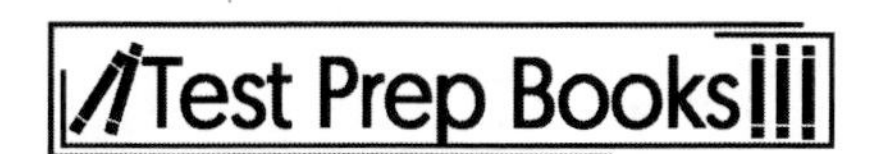

9. Which of the following would be the best choice for this sentence (reproduced below)?

(5) These advanced offices oftentimes require a Professional Engineering (PE) license which can be obtained through additional college courses, professional experience, and acceptable scores on the Fundamentals of Engineering (FE) and Professional Engineering (PE) standardized assessments.

a. NO CHANGE
b. These advanced positions oftentimes require acceptable scores on the Fundamentals of Engineering (FE) and Professional Engineering (PE) standardized assessments in order to achieve a Professional Engineering (PE) license. Additional college courses and professional experience help.
c. These advanced offices oftentimes require acceptable scores on the Fundamentals of Engineering (FE) and Professional Engineering (PE) standardized assessments to gain the Professional Engineering (PE) license which can be obtained through additional college courses, professional experience.
d. These advanced positions oftentimes require a Professional Engineering (PE) license which is obtained by acceptable scores on the Fundamentals of Engineering (FE) and Professional Engineering (PE) standardized assessments. Further education and professional experience can help prepare for the assessments.

10. Which of the following would be the best choice for this sentence (reproduced below)?

(6) Once the job begins, this lines of work requires critical thinking, business skills, problem solving, and creativity.

e. NO CHANGE
f. Once the job begins, this line of work
g. Once the job begins, these line of work
h. Once the job begin, this line of work

11. Which of the following would be the best choice for this sentence (reproduced below)?

This level of (7) expertise allows aircraft engineers to apply mathematical equation and scientific processes to aeronautical and aerospace issues or inventions.

a. NO CHANGE
b. expertis
c. expirtise
d. excpertise

12. Which of the following would be the best choice for this sentence (reproduced below)?

This level of expertise (8) allows aircraft engineers to apply mathematical equation and scientific processes to aeronautical and aerospace issues or inventions.

e. NO CHANGE
f. Inhibits
g. Requires
h. Should

13. Which of the following would be the best choice for this sentence (reproduced below)?

This level of expertise allows aircraft engineers to (9) apply mathematical equation and scientific processes to aeronautical and aerospace issues or inventions.

a. NO CHANGE
b. apply mathematical equations and scientific process
c. apply mathematical equation and scientific process
d. apply mathematical equations and scientific processes

14. Which of the following would be the best choice for this sentence (reproduced below)?

(10) For example, aircraft engineers may test, design, and construct flying vessels such as airplanes, space shuttles, and missile weapons.

e. NO CHANGE
f. Therefore,
g. However,
h. Furthermore,

15. Which of the following would be the best choice for this sentence (reproduced below)?

(11) In May 2015, the United States Bureau of Labor Statistics (BLS) reported that the median annual salary of aircraft engineers was $107,830.

a. NO CHANGE
b. May of 2015, the United States Bureau of Labor Statistics (BLS) reported that the median annual salary of aircraft engineers was $107,830.
c. In May of 2015 the United States Bureau of Labor Statistics (BLS) reported that the median annual salary of aircraft engineers was $107,830.
d. In May, 2015, the United States Bureau of Labor Statistics (BLS) reported that the median annual salary of aircraft engineers was $107,830.

16. Which of the following would be the best choice for this sentence (reproduced below)?

(12) Conversely, employment opportunities for aircraft engineers are projected to decrease by 2 percent by 2024.

e. NO CHANGE
f. Similarly,
g. In other words,
h. Accordingly,

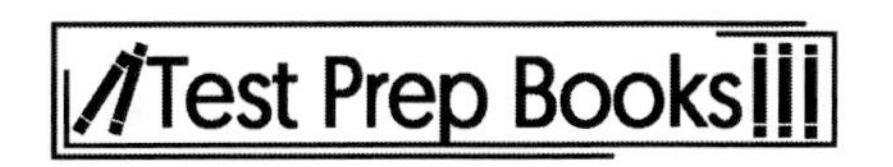

17. Which of the following would be the best choice for this sentence (reproduced below)?

Conversely, (13) employment opportunities for aircraft engineers are projected to decrease by 2 percent by 2024.

a. NO CHANGE
b. employment opportunities for aircraft engineers will be projected to decrease by 2 percent by 2024.
c. employment opportunities for aircraft engineers is projected to decrease by 2 percent by 2024.
d. employment opportunities for aircraft engineers was projected to decrease by 2 percent by 2024.

18. Which of the following would be the best choice for this sentence (reproduced below)?

(14) Nevertheless aircraft engineers who know how to utilize modeling and simulation programs, fluid dynamic software, and robotic engineering tools is projected to remain the most employable.

e. NO CHANGE
f. Nevertheless; aircraft engineers who know how to utilize
g. Nevertheless, aircraft engineers who know how to utilize
h. Nevertheless—aircraft engineers who know how to utilize

19. Which of the following would be the best choice for this sentence (reproduced below)?

Nevertheless aircraft engineers who know how to utilize modeling and simulation programs, fluid dynamic software, and robotic engineering tools (15) is projected to remain the most employable.

a. NO CHANGE
b. am projected to remain
c. was projected to remain
d. are projected to remain

Questions 20–25 are based on the following passage:

On September 11th, 2001, a group of terrorists hijacked four American airplanes. The terrorists crashed the planes into the World Trade Center in New York City, the Pentagon in Washington D.C., and a field in Pennsylvania. Nearly 3,000 people died during the attacks, which propelled the United States into a (16) "War on Terror".

About the Terrorists

(17) Terrorists commonly uses fear and violence to achieve political goals. The nineteen terrorists who orchestrated and implemented the attacks of September 11th were militants associated with al-Qaeda, an Islamic extremist group founded by Osama bin Laden, Abdullah Azzam, and others in the late 1980s. (18) Bin Laden orchestrated the attacks as a response to what he felt was American injustice against Islam and hatred towards Muslims. In his words, "Terrorism against America deserves to be praised."

Islam is the religion of Muslims, (19) who live mainly in south and southwest Asia and Sub-Saharan Africa. The majority of Muslims practice Islam peacefully. However, fractures in Islam have led to the growth of Islamic extremists who strictly oppose

Western influences. They seek to institute stringent Islamic law and destroy those who violate Islamic code.

In November 2002, bin Laden provided the explicit motives for the 9/11 terror attacks. According to this list, (20) Americas support of Israel, military presence in Saudi Arabia, and other anti-Muslim actions were the causes.

The Timeline of the Attacks

The morning of September 11 began like any other for most Americans. Then, at 8:45 a.m., a Boeing 767 plane (21) crashed into the north tower of the World Trade Center in New York City. Hundreds were instantly killed. Others were trapped on higher floors. The crash was initially thought to be a freak accident. When a second plane flew directly into the south tower eighteen minutes later, it was determined that America was under attack.

At 9:45 a.m., (22) slamming into the Pentagon was a third plane, America's military headquarters in Washington D.C. The jet fuel of this plane caused a major fire and partial building collapse that resulted in nearly 200 deaths. By 10:00 a.m., the south tower of the World Trade Center collapsed. Thirty minutes later, the north tower followed suit.

While this was happening, a fourth plane that departed from New Jersey, United Flight 93, was hijacked. The passengers learned of the attacks that occurred in New York and Washington D.C. and realized that they faced the same fate as the other planes that crashed. The passengers were determined to overpower the terrorists in an effort to prevent the deaths of additional innocent American citizens. Although the passengers were successful in (23) diverging the plane, it crashed in a western Pennsylvania field and killed everyone on board. The plane's final target remains uncertain, (24) but believed by many people was the fact that United Flight 93 was heading for the White House.

Heroes and Rescuers

(25) Close to 3,000 people died in the World Trade Center attacks. This figure includes 343 New York City firefighters and paramedics, 23 New York City police officers, and 37 Port Authority officers. Nevertheless, thousands of men and women in service worked (26) valiantly to evacuate the buildings, save trapped workers, extinguish infernos, uncover victims trapped in fallen rubble, and tend to nearly 10,000 injured individuals.

About 300 rescue dogs played a major role in the after-attack salvages. Working twelve-hour shifts, the dogs scoured the rubble and alerted paramedics when they found signs of life. While doing so, the dogs served as a source of comfort and therapy for the rescue teams.

Initial Impacts on America

The attacks of September 11, 2001 resulted in the immediate suspension of all air travel. No flights could take off from or land on American soil. (27) American airports and airspace closed to all national and international flights. Therefore, over five hundred

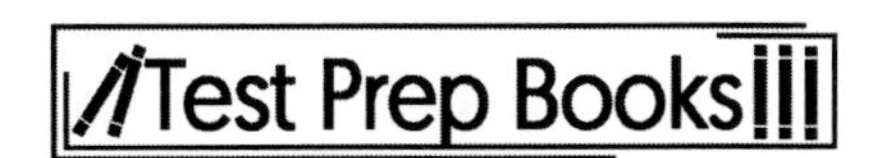

flights had to turn back or be redirected to other countries. Canada alone received 226 flights and thousands of stranded passengers. (28) Needless to say, as cancelled flights are rescheduled, air travel became backed up and chaotic for quite some time.

At the time of the attacks, George W. Bush was the president of the United States. President Bush announced that "We will make no distinction between the terrorists who committed these acts and those who harbor them." The rate of hate crimes against American Muslims spiked, despite President Bush's call for the country to treat them with respect.

Additionally, relief funds were quickly arranged. The funds were used to support families of the victims, orphaned children, and those with major injuries. In this way, the tragic event brought the citizens together through acts of service towards those directly impacted by the attack.

Long-Term Effects of the Attacks

Over the past fifteen years, the attacks of September 11th have transformed the United States' government, travel safety protocols, and international relations. Anti-terrorism legislation became a priority for many countries as law enforcement and intelligence agencies teamed up to find and defeat alleged terrorists.

Present George W. Bush announced a War on Terror. He (29) desired to bring bin Laden and al-Qaeda to justice and prevent future terrorist networks from gaining strength. The War in Afghanistan began in October of 2001 when the United States and British forces bombed al-Qaeda camps. (30) The Taliban, a group of fundamental Muslims who protected Osama bin Laden, was overthrown on December 9, 2001. However, the war continued in order to defeat insurgency campaigns in neighboring countries. Ten years later, the United State Navy SEALS killed Osama bin Laden in Pakistan. During 2014, the United States declared the end of its involvement in the War on Terror in Afghanistan.

Museums and memorials have since been erected to honor and remember the thousands of people who died during the September 11th attacks, including the brave rescue workers who gave their lives in the effort to help others.

20. Which of the following would be the best choice for this sentence (reproduced below)?

Nearly 3,000 people died during the attacks, which propelled the United States into a (16) "War on Terror".

e. NO CHANGE
f. "war on terror".
g. "war on terror."
h. "War on Terror."

21. Which of the following would be the best choice for this sentence (reproduced below)?

(17) Terrorists commonly uses fear and violence to achieve political goals.

a. NO CHANGE
b. Terrorist's commonly use fear and violence to achieve political goals.
c. Terrorists commonly use fear and violence to achieve political goals.
d. Terrorists commonly use fear and violence to achieves political goals.

22. Which of the following would be the best choice for this sentence (reproduced below)?

(18) Bin Laden orchestrated the attacks as a response to what he felt was American injustice against Islam and hatred towards Muslims.

e. NO CHANGE
f. Bin Laden orchestrated the attacks as a response to what he felt was American injustice against Islam, and hatred towards Muslims.
g. Bin Laden orchestrated the attacks, as a response to what he felt was American injustice against Islam and hatred towards Muslims.
h. Bin Laden orchestrated the attacks as responding to what he felt was American injustice against Islam and hatred towards Muslims.

23. Which of the following would be the best choice for this sentence (reproduced below)?

Islam is the religion of Muslims, (19) who live mainly in south and southwest Asia and Sub-Saharan Africa.

a. NO CHANGE
b. who live mainly in the South and Southwest Asia
c. who live mainly in the south and Southwest Asia
d. who live mainly in the south and southwest asia

24. Which of the following would be the best choice for this sentence (reproduced below)?

According to this list, (20) Americas support of Israel, military presence in Saudi Arabia, and other anti-Muslim actions were the causes.

e. NO CHANGE
f. America's support of israel,
g. Americas support of Israel
h. America's support of Israel,

25. Which of the following would be the best choice for this sentence (reproduced below)?

Then, at 8:45 a.m., a Boeing 767 plane (21) crashed into the north tower of the World Trade Center in New York City.

a. NO CHANGE
b. crashes into the north tower of the World Trade Center
c. crashing into the north tower of the World Trade Center
d. crash into the north tower of the World Trade Center

Reading Comprehension

Read the following poem and answer questions 26 & 27.

Two roads diverged in a yellow wood,
And sorry I could not travel both
And be one traveler, long I stood
And looked down one as far as I could
To where it bent in the undergrowth; *5*

Then took the other, as just as fair,
And having perhaps the better claim,
Because it was grassy and wanted wear;
Though as for that the passing there
Had worn them really about the same, *10*

And both that morning equally lay
In leaves no step had trodden black.
Oh, I kept the first for another day!
Yet knowing how way leads on to way,
I doubted if I should ever come back. *15*

I shall be telling this with a sigh
Somewhere ages and ages hence:
Two roads diverged in a wood, and I—
I took the one less traveled by,
And that has made all the difference. *20*

"The Road Not Taken" by Robert Frost

26. Which option best expresses the symbolic meaning of the "road" and the overall theme?
    e. A divergent spot where the traveler had to choose the correct path to his destination
    f. A choice between good and evil that the traveler needs to make
    g. The traveler's struggle between his lost love and his future prospects
    h. Life's journey and the choices with which humans are faced

27. Which line best contributes to the idea that the second path was *less traveled*?
    a. 5
    b. 4
    c. 8
    d. 10

Questions 28–33 are based upon the following passage:

"Did you ever come across a protégé of his—one Hyde?" He asked.

"Hyde?" repeated Lanyon. "No. Never heard of him. Since my time."

That was the amount of information that the lawyer carried back with him to the great, dark bed on which he tossed to and fro until the small hours of the morning began to grow large. It was a night of little ease to his toiling mind, toiling in mere darkness and besieged by questions.

Six o'clock struck on the bells of the church that was so conveniently near to Mr. Utterson's dwelling, and still he was digging at the problem. Hitherto it had touched him on the intellectual side alone; but now his imagination also was engaged, or rather enslaved; and as he lay and tossed in the gross darkness of the night in the curtained room, Mr. Enfield's tale went by before his mind in a scroll of lighted pictures. He would be aware of the great field of lamps in a nocturnal city; then of the figure of a man walking swiftly; then of a child running from the doctor's; and then these met, and that human Juggernaut trod the child down and passed on regardless of her screams. Or else he would see a room in a rich house, where his friend lay asleep, dreaming and smiling at his dreams; and then the door of that room would be opened, the curtains of the bed plucked apart, the sleeper recalled, and, lo! There would stand by his side a figure to whom power was given, and even at that dead hour he must rise and do its bidding. The figure in these two phrases haunted the lawyer all night; and if at anytime he dozed over, it was but to see it glide more stealthily through sleeping houses, or move the more swiftly, and still the more smoothly, even to dizziness, through wider labyrinths of lamplighted city, and at every street corner crush a child and leave her screaming. And still the figure had no face by which he might know it; even in his dreams it had no face, or one that baffled him and melted before his eyes; and thus there it was that there sprung up and grew apace in the lawyer's mind a singularly strong, almost an inordinate, curiosity to behold the features of the real Mr. Hyde. If he could but once set eyes on him, he thought the mystery would lighten and perhaps roll altogether away, as was the habit of mysterious things when well examined. He might see a reason for his friend's strange preference or bondage, and even for the startling clauses of the will. And at least it would be a face worth seeing: the face of a man who was without bowels of mercy: a face which had but to show itself to raise up, in the mind of the unimpressionable Enfield, a spirit of enduring hatred.

From that time forward, Mr. Utterson began to haunt the door in the by-street of shops. In the morning before office hours, at noon when business was plenty of time scarce, at night under the face of the full city moon, by all lights and at all hours of solitude or concourse, the lawyer was to be found on his chosen post.

"If he be Mr. Hyde," he had thought, "I should be Mr. Seek."

Excerpt from The Strange Case of Dr. Jekyll and Mr. Hyde by Robert Louis Stevenson

28. What is the purpose of the use of repetition in the following passage?

    It was a night of little ease to his toiling mind, toiling in mere darkness and besieged by questions.

    e. It serves as a demonstration of the mental state of Mr. Lanyon.
    f. It is reminiscent of the church bells that are mentioned in the story.
    g. It mimics Mr. Utterson's ambivalence.
    h. It emphasizes Mr. Utterson's anguish in failing to identify Hyde's whereabouts.

29. What is the setting of the story in this passage?
    a. In the city
    b. On the countryside
    c. In a jail
    d. In a mental health facility

30. What can one infer about the meaning of the word "Juggernaut" from the author's use of it in the passage?
    e. It is an apparition that appears at daybreak.
    f. It scares children.
    g. It is associated with space travel.
    h. Mr. Utterson finds it soothing.

31. What is the definition of the word *haunt* in the following passage?

    From that time forward, Mr. Utterson began to haunt the door in the by-street of shops. In the morning before office hours, at noon when business was plenty of time scarce, at night under the face of the full city moon, by all lights and at all hours of solitude or concourse, the lawyer was to be found on his chosen post.

    a. To levitate
    b. To constantly visit
    c. To terrorize
    d. To daunt

32. The phrase *labyrinths of lamplighted city* contains an example of what?
    e. Hyperbole
    f. Simile
    g. Juxtaposition
    h. Alliteration

33. What can one reasonably conclude from the final comment of this passage?

    "If he be Mr. Hyde," he had thought, "I should be Mr. Seek."

    a. The speaker is considering a name change.
    b. The speaker is experiencing an identity crisis.
    c. The speaker has mistakenly been looking for the wrong person.
    d. The speaker intends to continue to look for Hyde.

Questions 34–39 are based upon the following passage:

My gentleness and good behaviour had gained so far on the emperor and his court, and indeed upon the army and people in general, that I began to conceive hopes of getting my liberty in a short time. I took all possible methods to cultivate this favourable disposition. The natives came, by degrees, to be less apprehensive of any danger from me. I would sometimes lie down, and let five or six of them dance on my hand; and at last the boys and girls would venture to come and play at hide-and-seek in my hair. I had now made a good progress in understanding and speaking the language. The emperor had a mind one day to entertain me with several of the country shows, wherein they exceed all nations I have known, both for dexterity and magnificence. I was diverted with none so much as that of the rope-dancers, performed upon a slender white thread, extended about two feet, and twelve inches from the ground. Upon which I shall desire liberty, with the reader's patience, to enlarge a little.

This diversion is only practised by those persons who are candidates for great employments, and high favour at court. They are trained in this art from their youth, and are not always of noble birth, or liberal education. When a great office is vacant, either by death or disgrace (which often happens,) five or six of those candidates petition the emperor to entertain his majesty and the court with a dance on the rope; and whoever jumps the highest, without falling, succeeds in the office. Very often the chief ministers themselves are commanded to show their skill, and to convince the emperor that they have not lost their faculty. Flimnap, the treasurer, is allowed to cut a caper on the straight rope, at least an inch higher than any other lord in the whole empire. I have seen him do the summerset several times together, upon a trencher fixed on a rope which is no thicker than a common packthread in England. My friend Reldresal, principal secretary for private affairs, is, in my opinion, if I am not partial, the second after the treasurer; the rest of the great officers are much upon a par.

Excerpt from Gulliver's Travels into Several Remote Nations of the World by Jonathan Swift

34. Which of the following statements best summarizes the central purpose of this text?
   e. Gulliver details his fondness for the archaic yet interesting practices of his captors.
   f. Gulliver conjectures about the intentions of the aristocratic sector of society.
   g. Gulliver becomes acquainted with the people and practices of his new surroundings.
   h. Gulliver's differences cause him to become penitent around new acquaintances.

35. What is the word *principal* referring to in the following text?

   My friend Reldresal, principal secretary for private affairs, is, in my opinion, if I am not partial, the second after the treasurer; the rest of the great officers are much upon a par.

   a. Primary or chief
   b. An acolyte
   c. An individual who provides nurturing
   d. One in a subordinate position

36. What can the reader infer from this passage?

> I would sometimes lie down, and let five or six of them dance on my hand; and at last the boys and girls would venture to come and play at hide-and-seek in my hair.

e. The children tortured Gulliver.
f. Gulliver traveled because he wanted to meet new people.
g. Gulliver is considerably larger than the children who are playing around him.
h. Gulliver has a genuine love and enthusiasm for people of all sizes.

37. What is the significance of the word *mind* in the following passage?

> The emperor had a mind one day to entertain me with several of the country shows, wherein they exceed all nations I have known, both for dexterity and magnificence.

a. The ability to think
b. A collective vote
c. A definitive decision
d. A mythological question

38. Which of the following assertions does NOT support the fact that games are a commonplace event in this culture?

e. My gentleness and good behavior . . . short time.
f. They are trained in this art from their youth . . . liberal education.
g. Very often the chief ministers themselves are commanded to show their skill . . . not lost their faculty.
h. Flimnap, the treasurer, is allowed to cut a caper on the straight rope . . . higher than any other lord in the whole empire.

39. How do Flimnap and Reldresal demonstrate the community's emphasis on physical strength and leadership abilities?

a. Only children used Gulliver's hands as a playground.
b. The two men who exhibited superior abilities held prominent positions in the community.
c. Only common townspeople, not leaders, walk the straight rope.
d. No one could jump higher than Gulliver.

Questions 40–45 are based upon the following passage:

> Four score and seven years ago our fathers brought forth on this continent, a new nation, conceived in liberty, and dedicated to the proposition that all men are created equal.
>
> Now we are engaged in a great civil war, testing whether that nation, or any nation so conceived and so dedicated, can long endure. We are met on a great battlefield of that war. We have come to dedicate a portion of that field, as a final resting place for those who here gave their lives that this nation might live. It is altogether fitting and proper that we should do this.
>
> But, in a larger sense, we cannot dedicate—we cannot consecrate that we cannot hallow—this ground. The brave men, living and dead, who struggled here, have consecrated it, far above our poor power to add or detract. The world will little note, nor long remember what we say here, but it can never forget what they did here. It is

for us the living, rather, to be dedicated here to the unfinished work which they who fought here have thus far so nobly advanced. It is rather for us to be here and dedicated to the great task remaining before us—that from these honored dead we take increased devotion to that cause for which they gave the last full measure of devotion—that we here highly resolve that these dead shall not have died in vain—that this nation, under God, shall have a new birth of freedom—and that government of people, by the people, for the people, shall not perish from the earth.

Abraham Lincoln's Address Delivered at the Dedication of the Cemetery at Gettysburg, November 19, 1863

40. The best description for the phrase *four score and seven years ago* is which of the following?
    e. A unit of measurement
    f. A period of time
    g. A literary movement
    h. A statement of political reform

41. What is the setting of this text?
    a. A battleship off of the coast of France
    b. A desert plain on the Sahara Desert
    c. A battlefield in North America
    d. The residence of Abraham Lincoln

42. Which war is Abraham Lincoln referring to in the following passage?

    Now we are engaged in a great civil war, testing whether that nation, or any nation so conceived and so dedicated, can long endure.

    e. World War I
    f. The War of the Spanish Succession
    g. World War II
    h. The American Civil War

43. What message is the author trying to convey through this address?
    a. The audience should perpetuate the ideals of freedom that the soldiers died fighting for.
    b. The audience should honor the dead by establishing an annual memorial service.
    c. The audience should form a militia that would overturn the current political structure.
    d. The audience should forget the lives that were lost and discredit the soldiers.

44. Which rhetorical device is being used in the following passage?

    ...we here highly resolve that these dead shall not have died in vain—that this nation, under God, shall have a new birth of freedom—and that government of people, by the people, for the people, shall not perish from the earth.

    e. Antimetabole
    f. Antiphrasis
    g. Anaphora
    h. Epiphora

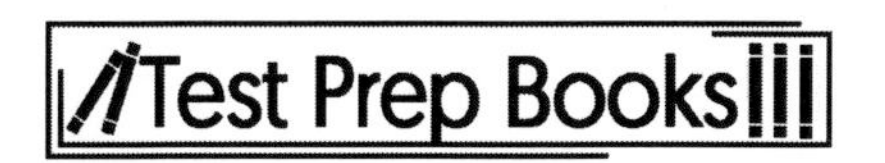

45. What is the effect of Lincoln's statement in the following passage?

 But, in a larger sense, we cannot dedicate—we cannot consecrate that we cannot hallow—this ground. The brave men, living and dead, who struggled here, have consecrated it, far above our poor power to add or detract.

a. His comparison emphasizes the great sacrifice of the soldiers who fought in the war.
b. His comparison serves as a remainder of the inadequacies of his audience.
c. His comparison serves as a catalyst for guilt and shame among audience members.
d. His comparison attempts to illuminate the great differences between soldiers and civilians.

Questions 46–50 are based on the following passage:

"MANKIND being originally equals in the order of creation, the equality could only be destroyed by some subsequent circumstance; the distinctions of rich, and poor, may in a great measure be accounted for, and that without having recourse to the harsh ill sounding names of oppression and avarice. Oppression is often the consequence, but seldom or never the means of riches; and though avarice will preserve a man from being necessitously poor, it generally makes him too timorous to be wealthy.

But there is another and greater distinction for which no truly natural or religious reason can be assigned, and that is, the distinction of men into KINGS and SUBJECTS. Male and female are the distinctions of nature, good and bad the distinctions of heaven; but how a race of men came into the world so exalted above the rest, and distinguished like some new species, is worth enquiring into, and whether they are the means of happiness or of misery to mankind.

In the early ages of the world, according to the scripture chronology, there were no kings; the consequence of which was there were no wars; it is the pride of kings which throw mankind into confusion. Holland without a king hath enjoyed more peace for this last century than any of the monarchical governments in Europe. Antiquity favors the same remark; for the quiet and rural lives of the first patriarchs hath a happy something in them, which vanishes away when we come to the history of Jewish royalty.

Government by kings was first introduced into the world by the Heathens, from whom the children of Israel copied the custom. It was the most prosperous invention the Devil ever set on foot for the promotion of idolatry. The Heathens paid divine honors to their deceased kings, and the Christian world hath improved on the plan by doing the same to their living ones. How impious is the title of sacred majesty applied to a worm, who in the midst of his splendor is crumbling into dust!

As the exalting one man so greatly above the rest cannot be justified on the equal rights of nature, so neither can it be defended on the authority of scripture; for the will of the Almighty, as declared by Gideon and the prophet Samuel, expressly disapproves of government by kings. All anti-monarchical parts of scripture have been very smoothly glossed over in monarchical governments, but they undoubtedly merit the attention of countries, which have their governments yet to form. "Render unto Caesar the things which are Caesar's" is the scripture doctrine of courts, yet it is no support of monarchical government, for the Jews at that time were without a king, and in a state of vassalage to the Romans.

Near three thousand years passed away from the Mosaic account of the creation, till the Jews under a national delusion requested a king. Till then their form of government (except in extraordinary cases, where the Almighty interposed) was a kind of republic administered by a judge and the elders of the tribes. Kings they had none, and it was held sinful to acknowledge any being under that title but the Lord of Hosts. And when a man seriously reflects on the idolatrous homage which is paid to the persons of Kings, he need not wonder, that the Almighty ever jealous of his honor, should disapprove of a form of government which so impiously invades the prerogative of heaven.

Excerpt From "Common Sense" by Thomas Paine

46. According to the passage, what role does avarice, or greed, play in poverty?
    e. It can make a man very wealthy
    f. It is the consequence of wealth
    g. Avarice can prevent a man from being poor, but too fearful to be very wealthy
    h. Avarice is what drives a person to be very wealthy

47. Of these distinctions, which does the author believe to be beyond natural or religious reason?
    a. Good and bad
    b. Male and female
    c. Human and animal
    d. King and subjects

48. According to the passage, what are the Heathens responsible for?
    e. Government by kings
    f. Quiet and rural lives of patriarchs
    g. Paying divine honors to their living kings
    h. Equal rights of nature

49. Which of the following best states Paine's rationale for the denouncement of monarchy?
    a. It is against the laws of nature
    b. It is against the equal rights of nature and is denounced in scripture
    c. Despite scripture, a monarchal government is unlawful
    d. Neither the law nor scripture denounce monarchy

50. Based on the passage, what is the best definition of the word *idolatrous*?
    e. Worshipping heroes
    f. Being deceitful
    g. Sinfulness
    h. Engaging in illegal activities

Questions 51–56 are based upon the following passage:

Fellow citizens—Pardon me, and allow me to ask, why am I called upon to speak here today? What have I, or those I represent, to do with your national independence? Are the great principles of political freedom and of natural justice, embodied in that Declaration of Independence, extended to us? And am I therefore called upon to bring our humble offering to the national altar, and to confess the benefits, and express devout gratitude for the blessings, resulting from your independence to us?

Would to God, both for your sakes and ours, ours that an affirmative answer could be truthfully returned to these questions! Then would my task be light, and my burden easy and delightful. For who is there so cold that a nation's sympathy could not warm him? Who so obdurate and dead to the claims of gratitude that would not thankfully acknowledge such priceless benefits? Who so stolid and selfish, that would not give his voice to swell the hallelujahs of a nation's jubilee, when the chains of servitude had been torn from his limbs? I am not that man. In a case like that, the dumb my eloquently speak, and the lame man leap as an hart.

But, such is not the state of the case. I say it with a sad sense of the disparity between us. I am not included within the pale of this glorious and anniversary. Oh pity! Your high independence only reveals the immeasurable distance between us. The blessings in which you this day rejoice, I do not enjoy in common. The rich inheritance of justice, liberty, prosperity, and independence, bequeathed by your fathers, is shared by *you*, not by *me*. This Fourth of July is *yours,* not *mine*. You may rejoice, *I* must mourn. To drag a man in fetters into the grand illuminated temple of liberty, and call upon him to join you in joyous anthems, were inhuman mockery and sacrilegious irony. Do you mean, citizens, to mock me, by asking me to speak today? If so there is a parallel to your conduct. And let me warn you that it is dangerous to copy the example of a nation whose crimes, towering up to heaven, were thrown down by the breath of the Almighty, burying that nation and irrecoverable ruin! I can today take up the plaintive lament of a peeled and woe-smitten people.

By the rivers of Babylon, there we sat down. Yea! We wept when we remembered Zion. We hanged our harps upon the willows in the midst thereof. For there, they that carried us away captive, required of us a song; and they who wasted us required of us mirth, saying, "Sing us one of the songs of Zion." How can we sing the Lord's song in a strange land? If I forget thee, O Jerusalem, let my right hand forget her cunning. If I do not remember thee, let my tongue cleave to the roof of my mouth.

Excerpt from "What to the Slave is the Fourth of July?" by Frederick Douglass, Rochester, New York July 5, 1852

51. What is the tone of the first paragraph of this passage?
   a. Exasperated
   b. Inclusive
   c. Contemplative
   d. Nonchalant

52. Which word CANNOT be used synonymously with the term *obdurate* as it is conveyed in the text below?

> Who so obdurate and dead to the claims of gratitude, that would not thankfully acknowledge such priceless benefits?

   e. Steadfast
   f. Stubborn
   g. Contented
   h. Unwavering

53. What is the central purpose of this text?
    a. To demonstrate the author's extensive knowledge of the Bible
    b. To address the hypocrisy of the Fourth of July holiday
    c. To convince wealthy landowners to adopt new holiday rituals
    d. To explain why minorities often relished the notion of segregation in government institutions

54. Which statement serves as evidence for the question above?
    e. By the rivers of Babylon...down.
    f. Fellow citizens...today.
    g. I can...woe-smitten people.
    h. The rich inheritance of justice...*not by me*.

55. The statement below features an example of which of the following literary devices?

    Oh pity! Your high independence only reveals the immeasurable distance between us.

    a. Assonance
    b. Parallelism
    c. Amplification
    d. Hyperbole

56. The speaker's use of biblical references, such as "rivers of Babylon" and the "songs of Zion," helps the reader to do all EXCEPT which of the following?
    e. Identify with the speaker using common text
    f. Convince the audience that injustices have been committed by referencing another group of people who have been previously affected by slavery
    g. Display the equivocation of the speaker and those that he represents
    h. Appeal to the listener's sense of humanity

Questions 57–62 are based upon the following passage:

> Three years ago, I think there were not many bird-lovers in the United States who believed it possible to prevent the total extinction of both egrets from our fauna. All the known rookeries accessible to plume-hunters had been totally destroyed. Two years ago, the secret discovery of several small, hidden colonies prompted William Dutcher, President of the National Association of Audubon Societies, and Mr. T. Gilbert Pearson, Secretary, to attempt the protection of those colonies. With a fund contributed for the purpose, wardens were hired and duly commissioned. As previously stated, one of those wardens was shot dead in cold blood by a plume hunter. The task of guarding swamp rookeries from the attacks of money-hungry desperadoes to whom the accursed plumes were worth their weight in gold, is a very chancy proceeding. There is now one warden in Florida who says that "before they get my rookery they will first have to get me."
>
> Thus far the protective work of the Audubon Association has been successful. Now there are twenty colonies, which contain all told, about 5,000 egrets and about 120,000 herons and ibises which are guarded by the Audubon wardens. One of the most important is on Bird Island, a mile out in Orange Lake, central Florida, and it is ably defended by Oscar E. Baynard. To-day, the plume hunters who do not dare to raid the guarded rookeries are trying to study out the lines of flight of the birds, to and from their feeding-grounds, and shoot them in transit. Their motto is—"Anything to beat the law, and get the plumes." It is there that the state of Florida should take part in the war.

> The success of this campaign is attested by the fact that last year a number of egrets were seen in eastern Massachusetts—for the first time in many years. And so to-day the question is, can the wardens continue to hold the plume-hunters at bay?

Excerpt from Our Vanishing Wildlife by William T. Hornaday

57. The author's use of first-person pronouns in the following text does NOT have which of the following effects?

> Three years ago, I think there were not many bird-lovers in the United States who believed it possible to prevent the total extinction of both egrets from our fauna.

a. The phrase *I think* acts as a sort of hedging, where the author's tone is less direct and/or absolute.
b. It allows the reader to more easily connect with the author.
c. It encourages the reader to empathize with the egrets.
d. It distances the reader from the text by overemphasizing the story.

58. What purpose does the quote serve at the end of the first paragraph?
e. The quote shows proof of a hunter threatening one of the wardens.
f. The quote lightens the mood by illustrating the colloquial language of the region.
g. The quote provides an example of a warden protecting one of the colonies.
h. The quote provides much needed comic relief in the form of a joke.

59. What is the meaning of the word *rookeries* in the following text?

> To-day, the plume hunters who do not dare to raid the guarded rookeries are trying to study out the lines of flight of the birds, to and from their feeding-grounds, and shoot them in transit.

a. Houses in a slum area
b. A place where hunters gather to trade tools
c. A place where wardens go to trade stories
d. A colony of breeding birds

60. What is on Bird Island?
e. Hunters selling plumes
f. An important bird colony
g. Bird Island Battle between the hunters and the wardens
h. An important egret with unique plumes

61. What is the main purpose of the passage?
a. To persuade the audience to act in preservation of the bird colonies
b. To show the effect hunting egrets has had on the environment
c. To argue that the preservation of bird colonies has had a negative impact on the environment
d. To demonstrate the success of the protective work of the Audubon Association

62. Why are hunters trying to study the lines of flight of the birds?
e. To study ornithology, one must know the lines of flight that birds take.
f. To help wardens preserve the lives of the birds
g. To have a better opportunity to hunt the birds
h. To build their homes under the lines of flight because they believe it brings good luck

Questions 63–67 are based upon the following passage:

Insects as a whole are preeminently creatures of the land and the air. This is shown not only by the possession of wings by a vast majority of the class, but by the mode of breathing to which reference has already been made, a system of branching air-tubes carrying atmospheric air with its combustion-supporting oxygen to all the insect's tissues. The air gains access to these tubes through a number of paired air-holes or spiracles, arranged segmentally in series.

It is of great interest to find that, nevertheless, a number of insects spend much of their time under water. This is true of not a few in the perfect winged state, as for example aquatic beetles and water-bugs ('boatmen' and 'scorpions') which have some way of protecting their spiracles when submerged, and, possessing usually the power of flight, can pass on occasion from pond or stream to upper air. But it is advisable in connection with our present subject to dwell especially on some insects that remain continually under water till they are ready to undergo their final moult and attain the winged state, which they pass entirely in the air. The preparatory instars of such insects are aquatic; the adult instar is aerial. All may-flies, dragon-flies, and caddis-flies, many beetles and two-winged flies, and a few moths thus divide their life-story between the water and the air. For the present we confine attention to the Stone-flies, the May-flies, and the Dragon-flies, three well-known orders of insects respectively called by systematists the Plecoptera, the Ephemeroptera and the Odonata.

In the case of many insects that have aquatic larvae, the latter are provided with some arrangement for enabling them to reach atmospheric air through the surface-film of the water. But the larva of a stone-fly, a dragon-fly, or a may-fly is adapted more completely than these for aquatic life; it can, by means of gills of some kind, breathe the air dissolved in water.

Excerpt from The Life-Story of Insects by Geo H. Carpenter

63. Which statement best details the central idea in this passage?
    a. It introduces certain insects that transition from water to air.
    b. It delves into entomology, especially where gills are concerned.
    c. It defines what constitutes as insects' breathing.
    d. It invites readers to have a hand in the preservation of insects.

64. Which definition most closely relates to the usage of the word *moult* in the passage?
    e. An adventure of sorts, especially underwater
    f. Mating act between two insects
    g. The act of shedding part or all of the outer shell
    h. Death of an organism that ends in a revival of life

65. What is the purpose of the first paragraph in relation to the second paragraph?
    a. The first paragraph serves as a cause, and the second paragraph serves as an effect.
    b. The first paragraph serves as a contrast to the second.
    c. The first paragraph is a description for the argument in the second paragraph.
    d. The first and second paragraphs are merely presented in a sequence.

66. What does the following sentence most nearly mean?

The preparatory instars of such insects are aquatic; the adult instar is aerial.

e. The volume of water is necessary to prep the insect for transition rather than the volume of the air.
f. The abdomen of the insect is designed like a star in the water as well as the air.
g. The stage of preparation in between molting is acted out in the water, while the last stage is in the air.
h. These insects breathe first in the water through gills, yet continue to use the same organs to breathe in the air.

67. Which of the statements reflect information that one could reasonably infer based on the author's tone?

a. The author's tone is persuasive and attempts to call the audience to action.
b. The author's tone is passionate due to excitement over the subject and personal narrative.
c. The author's tone is informative and exhibits interest in the subject of the study.
d. The author's tone is somber, depicting some anger at the state of insect larvae.

Math

1. If $6t + 4 = 16$, what is t?

a. 1
b. 2
c. 3
d. 4

2. The variable *y* is directly proportional to *x*. If $y = 3$ when $x = 5$, then what is *y* when $x = 20$?

e. 10
f. 12
g. 14
h. 16

3. A line passes through the point (1, 2) and crosses the *y*-axis at $y = 1$. Which of the following is an equation for this line?

a. $y = 2x$
b. $y = x + 1$
c. $x + y = 1$
d. $y = \frac{x}{2} - 2$

4. There are $4x + 1$ treats in each party favor bag. If a total of $60x + 15$ treats are distributed, how many bags are given out?

e. 15
f. 16
g. 20
h. 22

5. An accounting firm charted its income on the following pie graph. If the total income for the year was $500,000, how much of the income was received from Audit and Taxation Services?

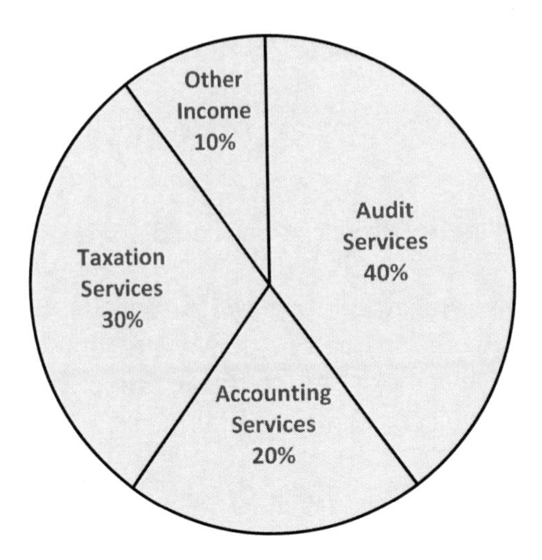

a. $200,000
b. $350,000
c. $150,000
d. $300,000

6. What are the roots of $x^2 + x - 2$?
e. 1 and -2
f. -1 and 2
g. 2 and -2
h. 9 and 13

7. What is the *y*-intercept of $y = x^{5/3} + (x - 3)(x + 1)$?
a. 3.5
b. 7.6
c. -3
d. -15.1

8. The phone bill is calculated each month using the equation $c = 50g + 75$. The cost of the phone bill per month is represented by c, and g represents the gigabytes of data used that month. What is the value and interpretation of the slope of this equation?
e. 75 dollars per day
f. 75 gigabytes per day
g. 50 dollars per day
h. 50 dollars per gigabyte

9. $(4x^2y^4)^{\frac{3}{2}}$ can be simplified to which of the following?

a. $8x^3y^6$

b. $4x^{\frac{5}{2}}y$

c. $4xy$

d. $32x^{\frac{7}{2}}y^{\frac{11}{2}}$

10. If $\sqrt{1+x} = 4$, what is x?

e. 10
f. 15
g. 20
h. 25

11. Suppose $\frac{x+2}{x} = 2$. What is x?

a. -1
b. 0
c. 2
d. 4

12. Which graph will be a line parallel to the graph of $y = 3x - 2$?

e. $2y - 6x = 2$
f. $y - 4x = 4$
g. $3y = x - 2$
h. $2x - 2y = 2$

13. A rectangle has a length that is 5 feet longer than three times its width. If the perimeter is 90 feet, what is the length in feet?

a. 10
b. 20
c. 25
d. 35

14. Five students take a test. The scores of the first four students are 80, 85, 75, and 60. If the median score is 80, which of the following could NOT be the score of the fifth student?

e. 60
f. 80
g. 85
h. 100

15. In an office, there are 50 workers. A total of 60% of the workers are women, and the chances of a woman wearing a skirt is 50%. If no men wear skirts, how many workers are wearing skirts?

a. 12
b. 15
c. 16
d. 20

16. Ten students take a test. Five students get a 50. Four students get a 70. If the average score is 55, what was the last student's score?

e. 20
f. 40
g. 50
h. 60

17. A company invests \$50,000 in a building where they can produce saws. If the cost of producing one saw is \$40, then which function expresses the amount of money the company pays? The variable *y* is the money paid and *x* is the number of saws produced.

a. $y = 50{,}000x + 40$
b. $y + 40 = x - 50{,}000$
c. $y = 40x - 50{,}000$
d. $y = 40x + 50{,}000$

18. A six-sided die is rolled. What is the probability that the roll is 1 or 2?

e. $\frac{1}{6}$
f. $\frac{1}{4}$
g. $\frac{1}{3}$
h. $\frac{1}{2}$

19. A line passes through the origin and through the point (-3, 4). What is the slope of the line?

a. $-\frac{4}{3}$
b. $-\frac{3}{4}$
c. $\frac{4}{3}$
d. $\frac{3}{4}$

20. An equilateral triangle has a perimeter of 18 feet. If a square whose sides have the same length as one side of the triangle is built, what will be the area of the square?

e. 6 square feet
f. 36 square feet
g. 256 square feet
h. 1000 square feet

21. Change $3\ \frac{3}{5}$ to a decimal.

a. 3.6
b. 4.67
c. 5.3
d. 0.28

22. The chart below shows the average car sales for the months of July through December for two different car dealers. What is the average number of cars sold in the given time period for Dealer 1?

e. 7
f. 11
g. 9
h. 8

23. The soccer team is selling donuts to raise money to buy new uniforms. For every box of donuts they sell, the team receives $3 towards their new uniforms. There are 15 people on the team. How many boxes does each player need to sell in order to raise $270 for their new uniforms?

a. 6
b. 30
c. 18
d. 5

24. What is the value of x in the diagram below?

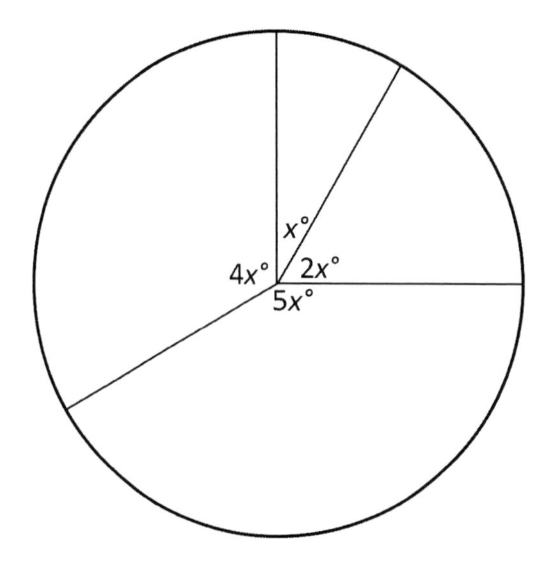

e. 60
f. 50
g. 30
h. 36

25. What is the probability of randomly picking the winner and runner-up from a race of 4 horses and distinguishing which is the winner?

a. $\frac{1}{4}$

b. $\frac{1}{2}$

c. $\frac{1}{16}$

d. $\frac{1}{12}$

26. A figure skater is facing north when she begins to spin to her right. She spins 2250 degrees. Which direction is she facing when she finishes her spin?

e. North
f. South
g. East
h. West

27. What is the measure of angle P?

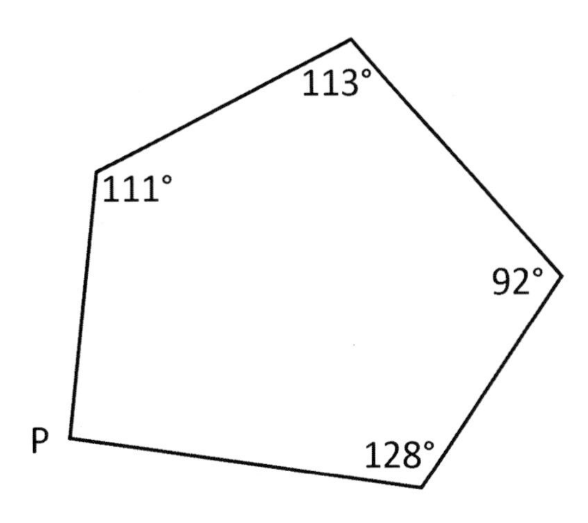

a. 84 degrees
b. 92 degrees
c. 96 degrees
d. 113 degrees

28. The expression $\frac{x-4}{x^2-6x+8}$ is undefined for what value(s) of x?
e. 4 and 2
f. -4 and -2
g. 2
h. 4

29. Nina has a jar where she puts her loose change at the end of each day. There are 13 quarters, 25 dimes, 18 nickels, and 30 pennies in the jar. If she chooses a coin at random, what is the probability that the coin will not be a penny or a dime?
a. 0.36
b. 0.64
c. 0.56
d. 0.34

30. The Cross family is planning a trip to Florida. They will be taking two cars for the trip. One car gets 18 miles to the gallon of gas. The other car gets 25 miles to the gallon. If the total trip to Florida is 450 miles, and the cost of gas is $2.49/gallon, how much will the gas cost for both cars to complete the trip?
e. $43.00
f. $44.82
g. $107.07
h. $32.33

31. Add and express in reduced form $\frac{14}{33}+\frac{10}{11}$.

a. $\frac{2}{11}$

b. $\frac{6}{11}$

c. $\frac{4}{3}$

d. $\frac{44}{33}$

32. 32 is 25% of what number?

e. 64
f. 128
g. 12.65
h. 8

33. If $3x-4+5x=8-10x$, what is the value of x?

a. 6
b. -6
c. 0.5
d. 0.67

34. Given the triangle below, find the value of x if $y=21$.

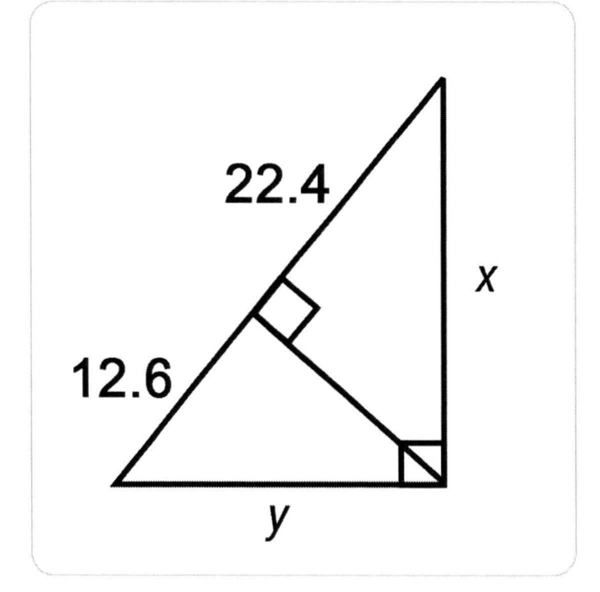

e. 35
f. 28
g. 25
h. 26

35. Of the given sets of coordinates below, which one lies on the line that is perpendicular to $y = 2x - 3$ and passes through the point $(0, 5)$?

a. $(2, 4)$
b. $(-2, 7)$
c. $(4, -3)$
d. $(-6, 10)$

36. Which of the following is a factor of both $x^2 + 4x + 4$ and $x^2 - x - 6$?

e. $x - 3$
f. $x + 2$
g. $x - 2$
h. $x + 3$

37. If $g(x) = x^3 - 3x^2 - 2x + 6$ and $f(x) = 2$, then what is $g(f(x))$?

a. -26
b. 6
c. $2x^3 - 6x^2 - 4x + 12$
d. -2

38. Which of the following is an equivalent measurement for 1.3 cm?

e. 0.13 m
f. 0.013 m
g. 0.13 mm
h. 0.013 mm

39. Divide $1{,}015 \div 1.4$.

a. 7,250
b. 725
c. 7.25
d. 72.50

40. What is the solution to the following system of equations?

$$x^2 - 2x + y = 8$$
$$x - y = -2$$

e. $(-2, 3)$
f. There is no solution.
g. $(-2, 0)$ $(1, 3)$
h. $(-2, 0)$ $(3, 5)$

41. Which of the following shows the correct result of simplifying the following expression:

$$(7n + 3n^3 + 3) + (8n + 5n^3 + 2n^4)$$

a. $9n^4 + 15n - 2$
b. $2n^4 + 5n^3 + 15n - 2$
c. $9n^4 + 8n^3 + 15n$
d. $2n^4 + 8n^3 + 15n + 3$

42. Multiply $1{,}987 \times 0.05$.

e. 9.935
f. 99.35
g. 993.5
h. 999.35

43. What is the product of the following expression?

$$(4x - 8)(5x^2 + x + 6)$$

a. $20x^3 - 36x^2 + 16x - 48$

b. $6x^3 - 41x^2 + 12x + 15$

c. $20x^3 + 11x^2 - 37x - 12$

d. $2x^3 - 11x^2 - 32x + 20$

44. For the following similar triangles, what are the values of x and y (rounded to one decimal place)?

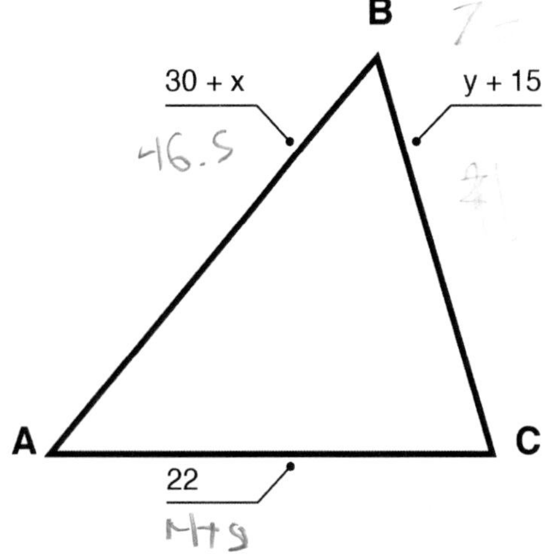

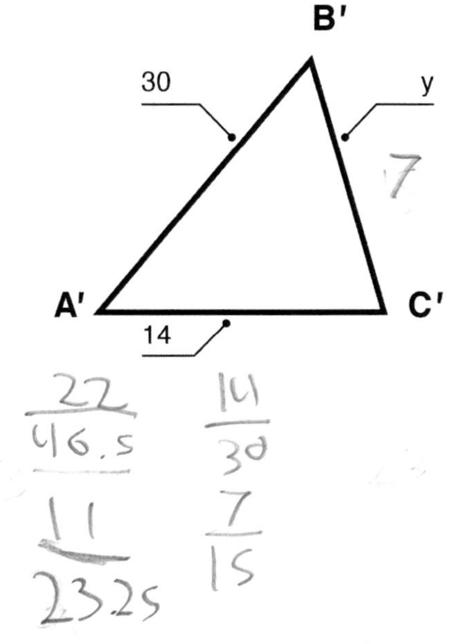

e. $x = 16.5, y = 25.1$
f. $x = 19.5, y = 24.1$
g. $x = 17.1, y = 26.3$
h. $x = 26.3, y = 17.1$

45. A solution needs 5 mL of saline for every 8 mL of medicine given. How much saline is needed for 45 mL of medicine?

a. $\frac{225}{8}$ mL

b. 72 mL

c. 28 mL

d. $\frac{45}{8}$ mL

46. If the volume of a sphere is 288π cubic meters, what are the radius and surface area of the same sphere?
    e. Radius 6 meters and surface area 144π square meters
    f. Radius 36 meters and surface area 144π square meters
    g. Radius 6 meters and surface area 12π square meters
    h. Radius 36 meters and surface area 12π square meters

47. The width of a rectangular house is 22 feet. What is the perimeter of this house if it has the same area as a house that is 33 feet wide and 50 feet long?
    a. 184 feet
    b. 200 feet
    c. 194 feet
    d. 206 feet

48. How much more area is covered by the rectangle than by the triangle?

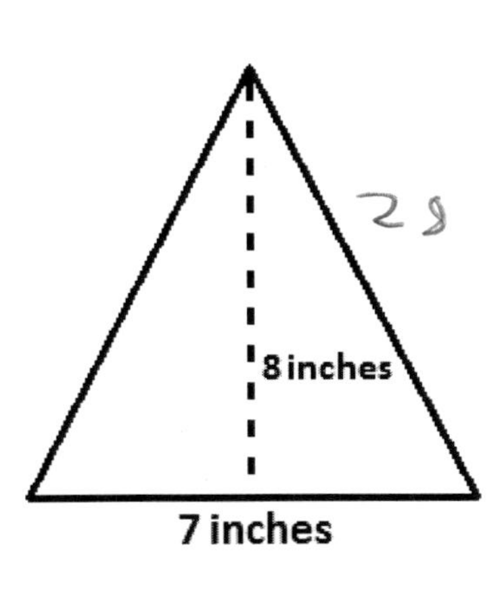

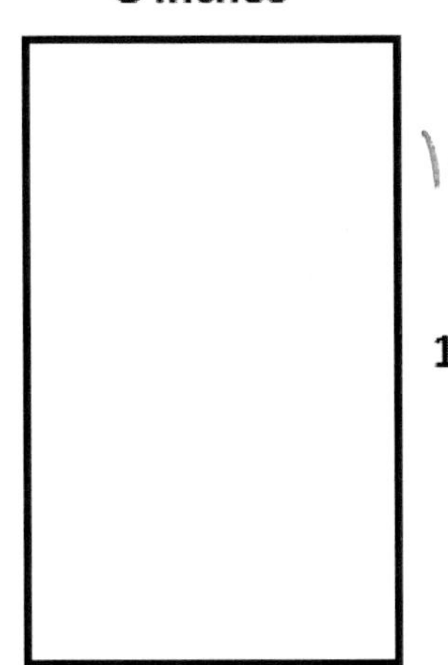

    e. 48
    f. 21
    g. 104
    h. 76

49. How will the following number be written in standard form: $(1 \times 10^4) + (3 \times 10^3) + (7 \times 10^1) + (8 \times 10^0)$
    a. 137
    b. 13,078
    c. 1,378
    d. 8,731

50. If $3x = 6y = -2z = 24$, then what does $4xy + z$ equal?
    e. 116
    f. 130
    g. 84
    h. 108

51. Johnny earns $2334.50 from his job each month. He pays $1437 for monthly expenses. Johnny is planning a vacation in 3 months that he estimates will cost $1750 total. How much will Johnny have left over from three months of saving once he pays for his vacation?

a. $948.50
b. $584.50
c. $852.50
d. $942.50

52. What is the value of the expression: $7^2 - 3 \times (4 + 2) + 15 \div 5$?

e. 12.2
f. 40.2
g. 34
h. 58.2

53. Which of the following is largest?

a. 0.45
b. 0.096
c. 0.3
d. 0.313

54. Which of the following is NOT a way to write 40 percent of *N*?

e. $(0.4)N$
f. $\frac{2}{5}N$
g. $40N$
h. $\frac{4N}{10}$

55. Which is closest to 17.8×9.9?

a. 140
b. 180
c. 200
d. 350

56. What is $\frac{420}{98}$ rounded to the nearest integer?

e. 4
f. 3
g. 5
h. 6

57. 16. A closet is filled with red, blue, and green shirts. If $\frac{1}{3}$ of the shirts are green and $\frac{2}{5}$ are red, what fraction of the shirts are blue?

a. $\frac{4}{15}$

b. $\frac{1}{5}$

c. $\frac{7}{15}$

d. $\frac{1}{2}$

58. If $2x + 6 = 20$, what is x?

e. 3
f. 4
g. 7
h. 9

59. Apples cost $2 each, while oranges cost $3 each. Maria purchased 10 fruits in total and spent $22. How many apples did she buy?

a. 5
b. 6
c. 7
d. 8

60. What are the polynomial roots of $x^2 + x - 2$?

e. 1 and -2
f. -1 and 2
g. 2 and -2
h. 9 and 13

61. Which measure for the center of a small sample set is most affected by outliers?

a. Mean
b. Median
c. Mode
d. Range

62. Given the value of a given stock at monthly intervals, which graph should be used to best represent the trend of the stock?

e. Box plot
f. Line plot
g. Scatter plot
h. Line graph

63. Gary is driving home to see his parents for Christmas. He travels at a constant speed of 60 miles per hour for a total of 350 miles. How many minutes will it take him to travel home if he takes a break for 10 minutes every 100 miles?

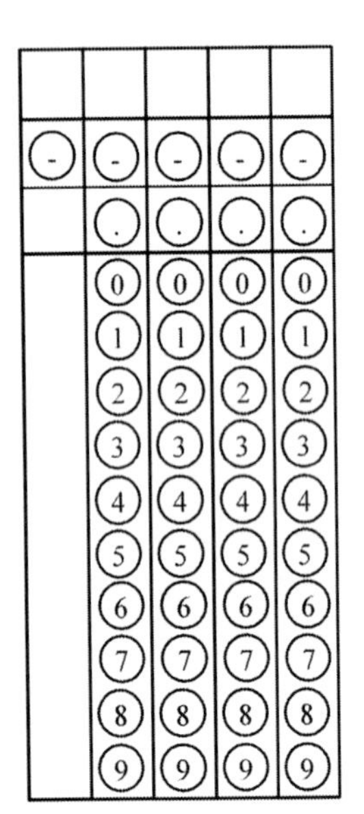

64. Kelly is selling cookies to raise money for the chorus. She has 500 cookies to sell. She sells 45% of the cookies to the sixth graders. At the second lunch, she sells 40% of what's left to the seventh graders. If she sells 60% of the remaining cookies to the eighth graders, how many cookies does Kelly have left at the end of all lunches?

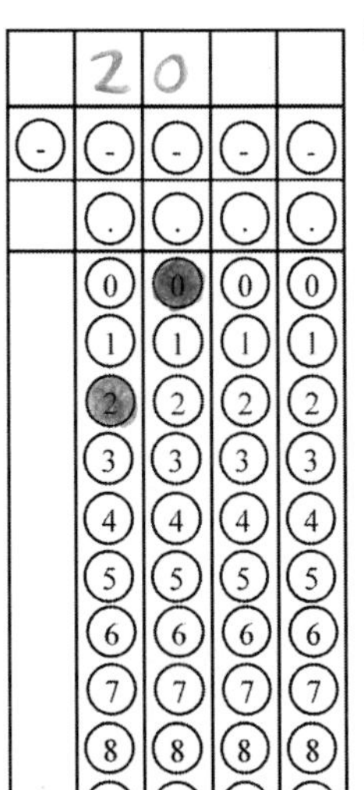

65. Find the value of x.

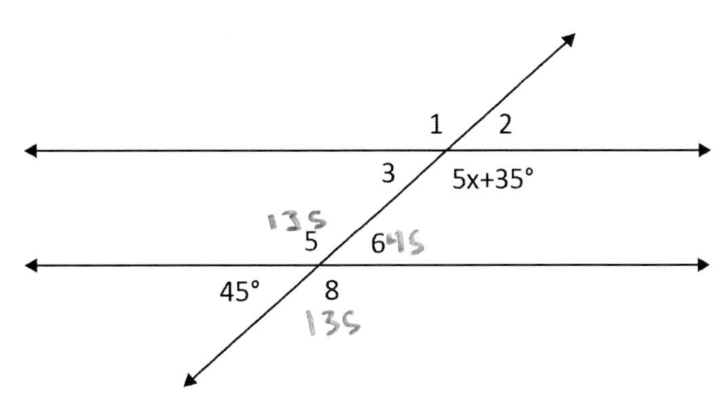

66. What is the value of the following expression?

$$\sqrt{8^2 + 6^2}$$

67. $864 \div 36 =$

Answer Explanations for Practice Test #1

Editing/Revising

1. D: The problem in the original sentence is that the second sentence is a dependent clause that cannot stand alone as a sentence; it must be attached to the main clause found in the first sentence. Because the main clause comes first, it does not need to be separated by a comma. However, if the dependent clause came first, then a comma would be necessary, which is why Choice *C* is incorrect. *A* and *B* also insert unnecessary commas into the sentence.

2. H: Choice *H* is correct because it uses correct parallel structure of plural nouns. *E* is incorrect because the word *shoe* should not be in singular form. Choice *F* is incorrect; semicolons are used in lists only if there is a list within a list. Choice *G* is incorrect because it pluralizes *makeup*, which is already in plural form.

3. C: The possessive form of the word "it" is "its." The contraction "it's" denotes "it is." Therefore, Choice *C* is correct. Choice *A* is incorrect because the end punctuation should be inside the quotes. Choice *B* is incorrect because removed the comma would create a run-on sentence. Distinguishes is the correct tense of the verb so Choice *D* is also incorrect.

4. F: Sentence 2 should be corrected to reflect the verb tense of the rest of the passage which is past tense. *Towards the end of his career, Carver returned to his first love of art.*

5. C: The next paragraph states that "These advanced offices oftentimes require a Professional Engineering (PE) license which can be obtained through additional college courses, professional experience, and acceptable scores on the Fundamentals of Engineering (FE) and Professional Engineering (PE) standardized assessments." Since the word *oftentimes* is used instead of *always*, Choice *C* is the best response.

6. G: The best answer is Choice *G*. Items in a list should be separated by a comma. Choice *E* is incorrect because there are no commas within the list to separate the items. Choice *F* is incorrect; a semicolon is used in a series only when a comma is present within the list itself. Choice *H* is incorrect because the conjunction *and* is missing before the word *calculus*.

7. A: The sentence is correct as-is. The words *one* and *bachelor* have apostrophe *-s* at the end because they show possession for the words that come after. The other answer choices do not indicate possession is being shown.

8. H: To begin, *of* is not required here. *Apprenticeship* is also more appropriate in this context than *apprentice opportunities*; *apprentice* describes an individual in an apprenticeship, not an apprenticeship itself. Both of these changes are needed, making Choice *H* the correct answer.

9. D: Choice *D* is correct because it breaks the section into coherent sentences and emphasizes the main point the author is trying to communicate: the PE license is required for some higher positions, it's obtained by scoring well on the two standardized assessments, and college and experience can be used to prepare for the assessments in order to gain the certification. The original sentence is a run-on and contains confusing information, so Choice *A* is incorrect. Choice *B* fixes the run-on aspect, but the sentence is indirect and awkward in construction. Choice *C* is incorrect for the same reason as Choice *B*, and it is a run on.

10. F: *Once the job begins, this line of work* is the best way to phrase this sentence. Choice *E* is incorrect because *lines* does not match up with *this*; it would instead match up with the word *these*. Choice *G* is incorrect; *these line* should say *this line*. Choice *H* is incorrect; *job begin* is faulty subject/verb agreement.

11. A: The word is spelled correctly as it is: *expertise*.

12. G: *Allows* is inappropriate because it does not stress what those in the position of aircraft engineers actually need to be able to do. *Requires* is the only alternative that fits because it actually describes necessary skills of the job.

13. D: The words *equations* and *processes* in this sentence should be plural. Choices *A, B,* and *C* have one or both words as singular, which is incorrect.

14. E: The correct response is Choice *E* because this statement's intent is to give examples as to how aircraft engineers apply mathematical equations and scientific processes towards aeronautical and aerospace issues and/or inventions. The answer is not *Therefore*, Choice *F,* or *Furthermore*, Choice *H,* because no causality is being made between ideas. Two items are neither being compared nor contrasted, so *However*, Choice *G,* is also not the correct answer.

15. A: No change is required. The comma is properly placed after the introductory phrase *In May of 2015*. Choice *B* is missing the word *In*. Choice *C* does not separate the introductory phrase from the rest of the sentence. Choice *D* places an extra comma prior to 2015.

16. E: The word *conversely* best demonstrates the opposite sentiments in this passage. Choice *F* is incorrect because it denotes agreement with the previous statement. Choice *G* is incorrect because the sentiment is not restated but opposed. Choice *H* is incorrect because the previous statement is not a cause for the sentence in question.

17. A: The correct answer is Choice *A*, *are projected*. The present tense *are* matches with the rest of the sentence. The verb *are* also matches with the plural *employment opportunities*. Choice *B* uses *will be projected*, which is incorrect because the statistic is being used as evidence, which demands a present or past tense verb. In this case it is present tense to maintain consistency. Choice *C* is incorrect because the singular verb *is* does not match with the plural subject *employment opportunities*. Choice *D* is incorrect because the past tense verb *were* does not maintain consistency with the present tense in the rest of the passage.

18. G: Choice *G* is the best answer because introductory words like "Nevertheless" are always succeeded by a comma.

19. D: The main subject and verb in this sentence are far apart from each other. The subject is *engineers* and the subject is *are projected*. Although there is a clause which interrupts the subject and the verb, they still must agree with each other.

20. H: The correct phrase should be "War on Terror." The phrase is capitalized because it was part of the campaign phrase that was launched by the U.S. government after September 11. Punctuation should always be used inside double quotes as well, making Choice *H* the best answer.

21. C: Terrorists commonly use fear and violence to achieve political goals. Choice *A* is incorrect because the subject *Terrorists* is plural while the verb *uses* is singular, so the subject and verb do not agree with each other. Choice *B* is incorrect because the word *Terrorist's* with the apostrophe -*s* shows possession, but the terrorists aren't in possession of anything in this sentence. Choice *D* is incorrect because the word *achieves* should be *achieve*.

22. E: No change is needed. Choices *F* and *G* utilize incorrect comma placements. Choice *H* utilizes an incorrect verb tense (*responding*).

23. B: The best answer Choice is *B*, *who live mainly in the South and Southwest Asia*. The directional terms *South Asia* and *Southwest Asia* are integral parts of a proper name and should therefore be capitalized.

24. H: This is the best answer choice because *America's* with the apostrophe *-s* shows possession of the word *support*, and *Israel* should be capitalized because it is a country and therefore a proper noun. Choice *E* does not show possession in the word *Americas*. Choice *F* does not capitalize the word *Israel*. Choice *G* does not show possession and does not include the necessary comma at the end of the phrase.

25. A: This sentence is correct as-is. The verb tense should be in the past—the other three answer choices either have a present or continuous verb tense, so these are incorrect.

Reading Comprehension

26. H: Choice *H* correctly summarizes Frost's theme of life's journey and the choices one makes. While Choice *E* can be seen as an interpretation, it is a literal one and is incorrect. Literal is not symbolic. Choice *F* presents the idea of good and evil as a theme, and the poem does not specify this struggle for the traveler. Choice *G* is a similarly incorrect answer. Love is not the theme.

27. C: Line 5 states, " Because it was grassy and wanted wear". The grass was taller and had not been trampled down by many travelers. The other choices are not lines that directly contribute to the idea that the second path was less traveled.

28. H: It emphasizes Mr. Utterson's anguish in failing to identify Hyde's whereabouts. Context clues indicate that Choice *H* is correct because the passage provides great detail of Mr. Utterson's feelings about locating Hyde. Choice *E* does not fit because there is no mention of Mr. Lanyon's mental state. Choice *F* is incorrect; although the text does make mention of bells, Choice *B* is not the *best* answer overall. Choice *G* is incorrect because the passage clearly states that Mr. Utterson was determined, not unsure.

29. A: In the city. The word *city* appears in the passage several times, thus establishing the location for the reader.

30. F: It scares children. The passage states that the Juggernaut causes the children to scream. Choices *E* and *H* don't apply because the text doesn't mention either of these instances specifically. Choice *G* is incorrect because there is nothing in the text that mentions space travel.

31. B: To constantly visit. The mention of *morning*, *noon*, and *night* make it clear that the word *haunt* refers to frequent appearances at various times. Choice *A* doesn't work because the text makes no mention of levitating. Choices *C* and *D* are not correct because the text makes mention of Mr. Utterson's anguish and disheartenment because of his failure to find Hyde but does not make mention of Mr. Utterson's feelings negatively affecting anyone else.

32. H: This is an example of alliteration. Choice *H* is the correct answer because of the repetition of the *L*-words. Hyperbole is an exaggeration, so Choice *E* doesn't work. No comparison is being made, so no simile or juxtaposition is being used, thus eliminating Choices *F* and *G*.

33. D: The speaker intends to continue to look for Hyde. Choices *A* and *B* are not possible answers because the text doesn't refer to any name changes or an identity crisis, despite Mr. Utterson's extreme obsession with finding Hyde. The text also makes no mention of a mistaken identity when referring to Hyde, so Choice *C* is also incorrect.

34. G: Gulliver becomes acquainted with the people and practices of his new surroundings. Choice *G* is the correct answer because it most extensively summarizes the entire passage. While Choices *E* and *F* are reasonable possibilities, they reference portions of Gulliver's experiences, not the whole. Choice *H* is incorrect because Gulliver doesn't express repentance or sorrow in this particular passage.

35. A: Principal refers to *chief* or *primary* within the context of this text. Choice *A* is the answer that most closely aligns with this definition. Choices *B* and *D* make reference to a helper or followers while Choice *C* doesn't meet the description of Reldresal from the passage.

36. G: One can reasonably infer that Gulliver is considerably larger than the children who were playing around him because multiple children could fit into his hand. Choice *F* is incorrect because there is no indication of stress in Gulliver's tone. Choices *E* and *H* aren't the best answer because though Gulliver seems fond of his new acquaintances, he didn't travel there with the intentions of meeting new people or to express a definite love for them in this particular portion of the text.

37. C: The emperor made a *definitive decision* to expose Gulliver to their native customs. In this instance, the word *mind* was not related to a vote, question, or cognitive ability.

38. E: Choice *E* is correct. This assertion does *not* support the fact that games are a commonplace event in this culture because it mentions conduct, not games. Choices *F*, *G*, and *H* are incorrect because these do support the fact that games were a commonplace event.

39. B: Choice *B* is the only option that mentions the correlation between physical ability and leadership positions. Choices *A* and *D* are unrelated to physical strength and leadership abilities. Choice *C* does not make a deduction that would lead to the correct answer—it only comments upon the abilities of common townspeople.

40. F: It denotes a period of time. It is apparent that Lincoln is referring to a period of time within the context of the passage because of how the sentence is structured with the word *ago*.

41. C: Lincoln's reference to *the brave men, living and dead, who struggled here,* proves that he is referring to a battlefield. Choices *A* and *B* are incorrect, as a *civil war* is mentioned and not a war with France or a war in the Sahara Desert. Choice *D* is incorrect because it does not make sense to consecrate a President's ground instead of a battlefield ground for soldiers who died during the American Civil War.

42. H: Abraham Lincoln is the former president of the United States, and he references a "civil war" during his address.

43. A: The audience should perpetuate the ideals of freedom that the soldiers died fighting for. Lincoln doesn't address any of the topics outlined in Choices *B*, *C*, or *D*. Therefore, Choice *A* is the correct answer.

44. H: Choice *H* is the correct answer because of the repetition of the word *people* at the end of the passage. Choice *E*, *antimetabole*, is the repetition of words in a succession. Choice *F*, *antiphrasis*, is a form of denial of an assertion in a text. Choice *G*, *anaphora*, is the repetition that occurs at the beginning of sentences.

45. A: Choice *A* is correct because Lincoln's intention was to memorialize the soldiers who had fallen as a result of war as well as celebrate those who had put their lives in danger for the sake of their country. Choices *B* and *D* are incorrect because Lincoln's speech was supposed to foster a sense of pride among the members of the audience while connecting them to the soldiers' experiences.

46. G: In lines 6 and 7, it is stated that avarice can prevent a man from being necessitously poor, but too timorous, or fearful, to achieve real wealth. According to the passage, avarice does not tend to make a person very wealthy. The passage states that oppression, not avarice, is the consequence of wealth. The passage does not state that avarice drives a person's desire to be wealthy.

47. D: Paine believes that the distinction that is beyond a natural or religious reason is between king and subjects. He states that the distinction between good and bad is made in heaven. The distinction between male and female is natural. He does not mention anything about the distinction between humans and animals.

48. E: The passage states that the Heathens were the first to introduce government by kings into the world. The quiet lives of patriarchs came before the Heathens introduced this type of government. It was Christians, not Heathens, who paid divine honors to living kings. Heathens honored deceased kings. Equal rights of nature are mentioned in the paragraph, but not in relation to the Heathens.

49. B: Paine asserts that a monarchy is against the equal rights of nature and cites several parts of scripture that also denounce it. He doesn't say it is against the laws of nature. Because he uses scripture to further his argument, it is not despite scripture that he denounces the monarchy. Paine addresses the law by saying the courts also do not support a monarchical government.

50. E: To be *idolatrous* is to worship idols or heroes, in this case, kings. It is not defined as being deceitful. While idolatry is considered a sin, it is an example of a sin, not a synonym for it. Idolatry may have been considered illegal in some cultures, but it is not a definition for the term.

51. A: The tone is exasperated. While contemplative is an option because of the inquisitive nature of the text, Choice *A* is correct because the speaker is frustrated by the thought of being included when he felt that the fellow members of his race were being excluded. The speaker is not nonchalant, nor accepting of the circumstances which he describes.

52. G: Choice *G*, *contented*, is the only word that has different meaning. Furthermore, the speaker expresses objection and disdain throughout the entire text.

53. B: To address the hypocrisy of the Fourth of July holiday. While the speaker makes biblical references, it is not the main focus of the passage, thus eliminating Choice *A* as an answer. The passage also makes no mention of wealthy landowners and doesn't speak of any positive response to the historical events, so Choices *C* and *D* are not correct.

54: H: Choice *H* is the correct answer because it clearly makes reference to justice being denied.

55: D: It is an example of hyperbole. Choices *A* and *B* are unrelated. Assonance is the repetition of sounds and commonly occurs in poetry. Parallelism refers to two statements that correlate in some

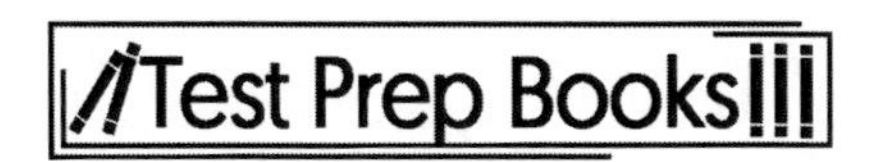

manner. Choice *C* is incorrect because amplification normally refers to clarification of meaning by broadening the sentence structure, while hyperbole refers to a phrase or statement that is being exaggerated.

56: G: Display the equivocation of the speaker and those that he represents. Choice *G* is correct because the speaker is clear about his intention and stance throughout the text. Choice *E* could be true, but the words "common text" is arguable. Choice *F* is also partially true, as another group of people affected by slavery are being referenced. However, the speaker is not trying to convince the audience that injustices have been committed, as it is already understood there have been injustices committed. Choice *H* is also close to the correct answer, but it is not the *best* answer choice possible.

57. D: The use of "I" could have all of the effects for the reader; it could serve to have a "hedging" effect, allow the reader to connect with the author in a more personal way, and cause the reader to empathize more with the egrets. However, it doesn't distance the reader from the text, thus eliminating Choice *D*.

58. G: The quote provides an example of a warden protecting one of the colonies. Choice *E* is incorrect because the speaker of the quote is a warden, not a hunter. Choice *F* is incorrect because the quote does not lighten the mood but shows the danger of the situation between the wardens and the hunters. Choice *H* is incorrect because there is no humor found in the quote.

59. D: A *rookery* is a colony of breeding birds. Although *rookery* could mean Choice *A*, houses in a slum area, it does not make sense in this context. Choices *B* and *C* are both incorrect, as this is not a place for hunters to trade tools or for wardens to trade stories.

60. F: An important bird colony. The previous sentence is describing "twenty colonies" of birds, so what follows should be a bird colony. Choice *E* may be true, but we have no evidence of this in the text. Choice *G* does touch on the tension between the hunters and wardens, but there is no official "Bird Island Battle" mentioned in the text. Choice *H* does not exist in the text.

61. D: To demonstrate the success of the protective work of the Audubon Association. The text mentions several different times how and why the association has been successful and gives examples to back this fact. Choice *A* is incorrect because although the article, in some instances, calls certain people to act, it is not the purpose of the entire passage. There is no way to tell if Choices *B* and *C* are correct, as they are not mentioned in the text.

62. G: To have a better opportunity to hunt the birds. Choice *E* might be true in a general sense, but it is not relevant to the context of the text. Choice *F* is incorrect because the hunters are not studying lines of flight to help wardens, but to hunt birds. Choice *H* is incorrect because nothing in the text mentions that hunters are trying to build homes underneath lines of flight of birds for good luck.

63. A: It introduces certain insects that transition from water to air. Choice *B* is incorrect because although the passage talks about gills, it is not the central idea of the passage. Choices *C* and *D* are incorrect because the passage does not "define" or "invite," but only serves as an introduction to stoneflies, dragonflies, and mayflies and their transition from water to air.

64. G: The act of shedding part or all of the outer shell. Choices *E*, *F*, and *H* are incorrect.

65. B: The first paragraph serves as a contrast to the second. Notice how the first paragraph goes into detail describing how insects are able to breathe air. The second paragraph acts as a contrast to the first by stating "[i]t is of great interest to find that, nevertheless, a number of insects spend much of their time under water." Watch for transition words such as "nevertheless" to help find what type of passage you're dealing with.

66. G: The stage of preparation in between molting is acted out in the water, while the last stage is in the air. Choices *E, F,* and *H* are all incorrect. *Instars* is the phase between two periods of molting, and the text explains when these transitions occur.

67. C: The author's tone is informative and exhibits interest in the subject of the study. Overall, the author presents us with information on the subject. One moment where personal interest is depicted is when the author states, "It is of great interest to find that, nevertheless, a number of insects spend much of their time under water."

Math

1. B: First, subtract 4 from each side. This yields $6t = 12$. Now, divide both sides by 6 to obtain $t = 2$.

2. F: To be directly proportional means that $y = mx$. If *x* is changed from 5 to 20, the value of *x* is multiplied by 4. Applying the same rule to the y-value, also multiply the value of *y* by 4. Therefore:

$$y = 12$$

3. B: From the slope-intercept form, $y = mx + b$, it is known that *b* is the *y*-intercept, which is 1. Compute the slope as $\frac{2-1}{1-0} = 1$, so the equation should be $y = x + 1$.

4. E: Each bag contributes $4x + 1$ treats. The total treats will be in the form $4nx + n$ where *n* is the total number of bags. The total is in the form $60x + 15$, from which it is known $n = 15$.

5. B: $350,000: Since the total income is $500,000, then a percentage of that can be found by multiplying the percent of Audit Services as a decimal, or 0.40, by the total of 500,000. This answer is found from the equation:

$$500000 \times 0.4 = 200000$$

The total income from Audit Services is $200,000.

For the income received from Taxation Services, the following equation can be used:

$$500000 \times 0.3 = 150000$$

The total income from Audit Services and Taxation Services is $150{,}000 + 200{,}000 = 350{,}000$.

Another way of approaching the problem is to calculate the easy percentage of 10% then multiply it by 7 because the total percentage for Audit and Taxation Services was 70%. 10% of 500,000 is 50,000. Then multiplying this number by 7 yields the same income of $350,000.

6. E: Finding the roots means finding the values of *x* when *y* is zero. The quadratic formula could be used, but in this case, it is possible to factor by hand, since the numbers -1 and 2 add to 1 and multiply to -2. So, factor $x^2 + x - 2 = (x - 1)(x + 2) = 0$, then set each factor equal to zero. Solving for each value gives the values *x* = 1 and *x* = -2.

7. C: To find the *y*-intercept, substitute zero for *x*, which gives us:

$$y = 0^{\frac{5}{3}} + (0 - 3)(0 + 1)$$

$$0 + (-3)(1) = -3$$

8. H: The slope from this equation is 50, and it is interpreted as the cost per gigabyte used. Since the *g*-value represents number of gigabytes and the equation is set equal to the cost in dollars, the slope relates these two values. For every gigabyte used on the phone, the bill goes up 50 dollars.

9. A: Simplify this to:

$$(4x^2y^4)^{\frac{3}{2}} = 4^{\frac{3}{2}}(x^2)^{\frac{3}{2}}(y^4)^{\frac{3}{2}}$$

$$4^{\frac{3}{2}} = (\sqrt{4})^3 = 2^3 = 8$$

For the other, recall that the exponents must be multiplied; this yields:

$$8x^{2\cdot\frac{3}{2}}y^{4\cdot\frac{3}{2}} = 8x^3y^6$$

10. F: Start by squaring both sides to get $1 + x = 16$. Then subtract 1 from both sides to get $x = 15$.

11. C: Multiply both sides by *x* to get $x + 2 = 2x$, which simplifies to $-x = -2$, or $x = 2$.

12. E: Parallel lines have the same slope. The slope of *G* can be seen to be 1/3 by dividing both sides by 3. The others are in standard form $Ax + By = C$, for which the slope is given by $\frac{-A}{B}$. The slope of *E* is 3, the slope of *F* is 4. The slope of *H* is 1.

13. D: Denote the width as w and the length as l. Then, $l = 3w + 5$. The perimeter is $2w + 2l = 90$. Substituting the first expression for l into the second equation yields:

$$2(3w + 5) + 2w = 90$$

$$6w + 10 + 2w = 90$$

$$8w = 80$$

$$w = 10$$

Putting this into the first equation, it yields:

$$l = 3(10) + 5 = 35$$

14. E: Lining up the given scores provides the following list: 60, 75, 80, 85, and one unknown. Because the median needs to be 80, it means 80 must be the middle data point out of these five. Therefore, the unknown data point must be the fourth or fifth data point, meaning it must be greater than or equal to 80. The only answer that fails to meet this condition is 60.

15. B: If 60% of 50 workers are women, then there are 30 women working in the office. If half of them are wearing skirts, then that means 15 women wear skirts. Since none of the men wear skirts, this means there are 15 people wearing skirts.

16. E: Let the unknown score be x. The average will be:

$$\frac{5 \times 50 + 4 \times 70 + x}{10} = \frac{530 + x}{10} = 55$$

Multiply both sides by 10 to get $530 + x = 550$, or $x = 20$.

17. D: For manufacturing costs, there is a linear relationship between the cost to the company and the number produced, with a *y*-intercept given by the base cost of acquiring the means of production, and a slope given by the cost to produce one unit. In this case, that base cost is $50,000, while the cost per unit is $40. So:

$$y = 40x + 50{,}000$$

18. G: A die has an equal chance for each outcome. Since it has six sides, each outcome has a probability of $\frac{1}{6}$. The chance of a 1 or a 2 is therefore:

$$\frac{1}{6} + \frac{1}{6} = \frac{1}{3}$$

19. A: The slope is given by:

$$m = \frac{y_2 - y_1}{x_2 - x_1} = \frac{0 - 4}{0 - (-3)} = -\frac{4}{3}$$

20. F: An equilateral triangle has three sides of equal length, so if the total perimeter is 18 feet, each side must be 6 feet long. A square with sides of 6 feet will have an area of $6^2 = 36$ square feet.

21. A: 3.6. Divide 3 by 5 to get 0.6 and add that to the whole number 3, to get 3.6. An alternative is to incorporate the whole number 3 earlier on by creating an improper fraction:

$$\frac{18}{5}$$

Then dividing 18 by 5 to get 3.6.

22. G: 9 Cars. The average is calculated by adding up each month's sales and dividing the sum by the total number of months in the time period. Dealer 1 sold 2 cars in July, 12 in August, 8 in September, 6 in October, 10 in November, and 15 in December. The sum of these sales is:

$$2 + 12 + 8 + 6 + 10 + 15 = 53 \text{ cars}$$

To find the average, this sum is divided by the total number of months, which is 6. When 53 is divided by 6, it yields 8.8333... Since cars are sold in whole numbers, the answer is rounded to 9 cars.

23. A: 6 boxes. The team needs a total of $270, and each box earns them $3. Therefore, the total number of boxes needed to be sold is $270 \div 3$, which is 90. With 15 people on the team, the total of 90 can be divided by 15, which equals 6. This means that each member of the team needs to sell 6 boxes for the team to raise enough money to buy new uniforms.

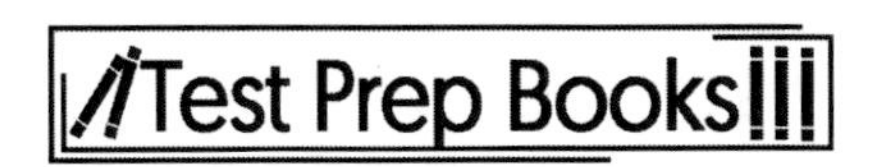

24. G: 30. A complete circle measures 360 degrees. This circle is broken up into 4 different parts with different measures for each part. Adding these parts should give a total of 360 degrees. The equation generated from this diagram is:

$$4x + 5x + x + 2x = 360$$

Collecting like terms gives the equation $12x = 360$, which can be solved by dividing by 12 to give $x = 30$. The value of x in the diagram is 30.

25. D: $\frac{1}{12}$. The probability of picking the winner of the race is $\frac{1}{4}$, or $\left(\frac{number\ of\ favorable\ outcomes}{number\ of\ total\ outcomes}\right)$. Assuming the winner was picked on the first selection, three horses remain from which to choose the runner-up (these are dependent events). Therefore, the probability of picking the runner-up is $\frac{1}{3}$. To determine the probability of multiple events, the probability of each event is multiplied:

$$\frac{1}{4} \times \frac{1}{3} = \frac{1}{12}$$

26. G: A full rotation is 360 degrees. Taking the total degrees that the figure skater spins and dividing by 360 yields 6.25. She spins 6 total times and then one quarter of a turn more. This quarter of a turn to her right means she ends up facing east.

27. C: The sum of all angles in a polygon with n sides is found by the expression $(n - 2) \times 180$. Since this polygon has 5 sides, then the total degrees of the interior angles can be found using the equation:

$$(5 - 2) \times 180 = 540$$

Adding up each of the given angles yields a total of:

$$111 + 113 + 92 + 128 = 444 \text{ degrees}$$

Taking the total of 540 degrees and subtracting the given sum of 444 degrees gives a missing value of 96 degrees for the measure of angle P.

28. E: The expression in the denominator can be factored into the two binomials:

$$(x - 4)(x - 2)$$

Once the expression is rewritten as $\frac{x-4}{(x-4)(x-2)}$, the values of $x = 4$ and $x = 2$ result in a denominator with a value of 0. Since 0 cannot be in the denominator of a fraction, the expression is undefined at the values of $x = 2, 4$.

29. A: The total number of coins in the jar is 86, which is the sum of all the coins. The probability of Nina choosing a coin other than a penny or a dime can be found by calculating the total of quarters and nickels. This total is 31. Taking 31 and dividing it by 86 gives the probability of choosing a coin that is not a penny or a dime. The decimal found from the fraction $\frac{31}{86}$ is 0.36.

30. G: For the first car, the trip will be 450 miles at 18 miles to the gallon. The total gallons needed for this car will be:

$$450 \div 18 = 25$$

For the second car, the trip will be 450 miles at 25 miles to the gallon, or $450 \div 25 = 18$, which will require 18 gallons of gas. Adding these two amounts of gas gives a total of 43 gallons of gas. If the gas costs $2.49 per gallon, the cost of the trip for both cars is:

$$43 \times \$2.49 = \$107.07$$

31. C: $\frac{4}{3}$

Set up the problem and find a common denominator for both fractions.

$$\frac{14}{33} + \frac{10}{11}$$

Multiply each fraction across by 1 to convert to a common denominator

$$\frac{14}{33} \times \frac{1}{1} + \frac{10}{11} \times \frac{3}{3}$$

Once over the same denominator, add across the top. The total is over the common denominator.

$$\frac{14 + 30}{33} = \frac{44}{33}$$

Reduce by dividing both numerator and denominator by 11.

$$\frac{44 \div 11}{33 \div 11} = \frac{4}{3}$$

32. F: 128

This question involves the percent formula.

$$\frac{32}{x} = \frac{25}{100}$$

We multiply the diagonal numbers, 32 and 100, to get 3,200. Dividing by the remaining number, 25, gives us 128.

The percent formula does not have to be used for a question like this. Since 25% is ¼ of 100, you know that 32 needs to be multiplied by 4, which yields 128.

33. D: The first step in solving this equation is to collect like terms on the left side of the equation. This yields the new equation:

$$-4 + 8x = 8 - 10x$$

The next step is to move the x-terms to one side by adding 10 to both sides, making the equation:

$$-4 + 18x = 8$$

Then the -4 can be moved to the right side of the equation to form:

$$18x = 12$$

Dividing both sides of the equation by 18 gives a value of 0.67, or $\frac{2}{3}$.

34. F: This triangle can be labeled as a right triangle because it has a right-angle measure in the corner. The Pythagorean Theorem can be used here to find the missing side lengths. The Pythagorean Theorem states that $a^2 + b^2 = c^2$, where a and b are side lengths and c is the hypotenuse. The hypotenuse, c, is equal to 35, and 1 side, a, is equal to 21. Plugging these values into the equation forms:

$$21^2 + b^2 = 35^2$$

Squaring both given numbers and subtracting them yields the equation:

$$b^2 = 784$$

Taking the square root of 784 gives a value of 28 for b. In the equation, b is the same as the missing side length x.

35. A: The first step is to find the equation of the line that is perpendicular to $y = 2x - 3$ and passes through the point $(0, 5)$. The slope of a perpendicular line is found by the negative reciprocal of 2, which is $-\frac{1}{2}$. The y-intercept is the value of y when $x = 0$, so the y-intercept is 5. The new equation is:

$$y = -\frac{1}{2}x + 5$$

In order to find which points lie on the new line, the values of x and y can be substituted into the equation to determine if they form a true statement. For A, the equation $4 = -\frac{1}{2}(2) + 5$ makes a true statement, so the point $(2, 4)$ lies on the lines.

For B, the equation $7 = -\frac{1}{2}(-2) + 5$ makes the statement $7 = 6$, which is not a true statement. Therefore, B is not a point that lies on the line. For C, the equation $-3 = -\frac{1}{2}(4) + 5$ becomes $-3 = 3$ which is not a true statement, so the point $(4, -3)$ is not on the line.

For the last point in D, the equation $10 = -\frac{1}{2}(-6) + 5$ makes the statement $10 = 8$. This is not a true statement, so the point $(-6, 10)$ does not lie on the line.

36. F: To factor $x^2 + 4x + 4$, the numbers needed are those that add to 4 and multiply to 4. Therefore, both numbers must be 2, and the expression factors to:

$$x^2 + 4x + 4 = (x + 2)^2$$

Similarly, the expression factors to $x^2 - x - 6 = (x - 3)(x + 2)$, so that they have $x + 2$ in common.

37. D: This problem involves a composition function, where one function is plugged into the other function. In this case, the $f(x)$ function is plugged into the $g(x)$ function for each x-value. The composition equation becomes:

$$g(f(x)) = 2^3 - 3(2^2) - 2(2) + 6$$

Simplifying the equation gives the answer:

$$g(f(x)) = 8 - 3(4) - 2(2) + 6$$

$$8 - 12 - 4 + 6 = -2$$

38. F: 100 cm is equal to 1 m. 1.3 divided by 100 is 0.013. Therefore, 1.3 cm is equal to 0.013 m. Because 1 cm is equal to 10 mm, 1.3 cm is equal to 13 mm.

39. B: 725

Set up the division problem.

$$1.4\overline{)1015}$$

Move the decimal over one place to the right in both numbers.

$$14\overline{)10150}$$

14 does not go into 1 or 10 but does go into 101 so start there.

```
      725
 14|10150
   - 98
     35
    - 28
      70
    - 70
       0
```

The result is 725.

40. H: This system of equations involves one quadratic function and one linear function, as seen from the degree of each equation. One way to solve this is through substitution. Solving for y in the second equation yields $y = x + 2$. Plugging this equation in for the y of the quadratic equation yields:

$$x^2 - 2x + x + 2 = 8$$

Simplifying the equation, it becomes:

$$x^2 - x + 2 = 8$$

Setting this equal to zero and factoring, it becomes:

$$x^2 - x - 6 = 0 = (x - 3)(x + 2)$$

Solving these two factors for x gives the zeros $x = 3, -2$. To find the y-value for the point, each number can be plugged in to either original equation. Solving each one for y yields the points $(3, 5)$ and $(-2, 0)$.

41. D: The expression is simplified by collecting like terms. Terms with the same variable and exponent are like terms, and their coefficients can be added.

42. F: 99.35. Set up the problem, with the larger number on top. Multiply as if there are no decimal places. Add the answer rows together. Count the number of decimal places that were in the original numbers (2).

Place the decimal in that many spots from the right for the final solution.

43. A: Finding the product means distributing one polynomial to the other so that each term in the first is multiplied by each term in the second. Then, like terms can be collected. Multiplying the factors yields the expression:

$$20x^3 + 4x^2 + 24x - 40x^2 - 8x - 48$$

Collecting like terms means adding the x^2 terms and adding the x terms. The final answer after simplifying the expression is:

$$20x^3 - 36x^2 + 16x - 48$$

44. G: Because the triangles are similar, the lengths of the corresponding sides are proportional. Therefore:

$$\frac{30 + x}{30} = \frac{22}{14} = \frac{y + 15}{y}$$

This results in the equation $14(30 + x) = 22 \times 30$ which, when solved, gives $x = 17.1$. The proportion also results in the equation $14(y + 15) = 22y$ which, when solved, gives $y = 26.3$.

45. A: Every 8 mL of medicine requires 5 mL. The 45 mL first needs to be split into portions of 8 mL. This results in $\frac{45}{8}$ portions. Each portion requires 5 mL. Therefore, the following is necessary:

$$\frac{45}{8} \times 5 = \frac{45 \times 5}{8} = \frac{225}{8} \text{ mL}$$

46. E: Because the volume of the given sphere is 288π cubic meters, this means:

$$\frac{4}{3}\pi r^3 = 288\pi$$

This equation is solved for r to obtain a radius of 6 meters. The formula for the surface area of a sphere is $4\pi r^2$, so if $r = 6$ in this formula, the surface area is 144π square meters.

47. C: First, find the area of the second house. The area is:

$$A = l \times w = 33 \times 50 = 1{,}650 \text{ square feet}$$

Then, use the area formula to determine what length gives the first house an area of 1,650 square feet. So:

$$1{,}650 = 22 \times l$$

$$l = \frac{1{,}650}{22} = 75 \text{ feet}$$

Then, use the formula for perimeter to get:

$$75 + 75 + 22 + 22 = 194 \text{ feet}$$

48. H: First, find the area of both figures. The area of the triangle is:

$$\frac{1}{2}(7) \times 8 = 28 \text{ square inches}$$

The area of the rectangle is:

$$13 \times 8 = 104 \text{ square inches}$$

To find how much more area is covered by the square, the following equation can be used:

$$104 - 28 = 76$$

49. B: 13,078. The power of 10 by which a digit is multiplied corresponds with the number of zeros following the digit when expressing its value in standard form. Therefore:

$$(1 \times 10^4) + (3 \times 10^3) + (7 \times 10^1) + (8 \times 10^0)$$

$$10{,}000 + 3{,}000 + 70 + 8 = 13{,}078$$

50. E: First solve for x, y, and z. So:

$$3x = 24$$

$$x = 8$$

$$6y = 24$$

$$y = 4$$

$$-2z = 24$$

$$z = -12$$

This means the equation would be $4(8)(4) + (-12)$, which equals 116.

51. D: First, subtract $1437 from $2334.50 to find Johnny's monthly savings; this equals $897.50. Then, multiply this amount by 3 to find out how much he will have in three months before he pays for his vacation; this equals $2692.50. Finally, subtract the cost of the vacation ($1750) from this total to find how much Johnny will have left: $942.50.

52. G: 34. When performing calculations consisting of more than one operation, the order of operations should be followed: *P*arenthesis, *E*xponents, *M*ultiplication/*D*ivision, *A*ddition/*S*ubtraction.

Parenthesis:

$$7^2 - 3 \times (4 + 2) + 15 \div 5$$

$$7^2 - 3 \times (6) + 15 \div 5$$

Exponents:

$$7^2 - 3 \times 6 + 15 \div 5$$

$$49 - 3 \times 6 + 15 \div 5$$

Multiplication/Division (from left to right):

$$49 - 3 \times 6 + 15 \div 5$$

$$49 - 18 + 3$$

Addition/Subtraction (from left to right):

$$49 - 18 + 3 = 34$$

53. A: Figure out which is largest by looking at the first non-zero digits. Choice *B*'s first non-zero digit is in the hundredths place. The other four all have non-zero digits in the tenths place, so it must be *A*, *C*, or *D*. Of these, *A* has the largest first non-zero digit.

54. G: 40*N* would be 4000% of *N*. It's possible to check that each of the others is actually 40% of *N*.

55. B: Instead of multiplying these out, the product can be estimated by using $18 \times 10 = 180$. The error here should be lower than 15, since it is rounded to the nearest integer, and the numbers add to something less than 30.

56. E: Dividing by 98 can be approximated by dividing by 100, which would mean shifting the decimal point of the numerator to the left by 2. The result is 4.2 and rounds to 4.

57. A: The total fraction taken up by green and red shirts will be:

$$\frac{1}{3} + \frac{2}{5} = \frac{5}{15} + \frac{6}{15} = \frac{11}{15}$$

The remaining fraction is:

$$1 - \frac{11}{15} = \frac{15}{15} - \frac{11}{15} = \frac{4}{15}$$

58. G: Begin by subtracting 6 from both sides to get $2x = 14$. Dividing both sides by 2 results in $x = 7$.

59. E: Let *a* be the number of apples and *o* the number of oranges. Then, the total cost is:

$$2a + 3o = 22$$

while it also known that $a + o = 10$. Using the knowledge of systems of equations, cancel the *o* variables by multiplying the second equation by -3.

This makes the equation:

$$-3a - 3o = -30$$

Adding this to the first equation, the b values cancel to get $-a = -8$, which simplifies to *a* = 8.

60. E: Finding the roots means finding the values of *x* when *y* is zero. The quadratic formula could be used, but in this case it is possible to factor by hand, since the numbers -1 and 2 add to 1 and multiply to -2. So, factor:

$$x^2 + x - 2 = (x - 1)(x + 2) = 0$$

then set each factor equal to zero. Solving for each value gives the values $x = 1$ and $x = -2$.

61. A: Mean. An outlier is a data value that is either far above or far below the majority of values in a sample set. The mean is the average of all the values in the set. In a small sample set, a very high or very low number could drastically change the average of the data points. Outliers will have no more of an effect on the median (the middle value when arranged from lowest to highest) than any other value above or below the median. If the same outlier does not repeat, outliers will have no effect on the mode (value that repeats most often).

62. H: Line graph. The scenario involves data consisting of two variables, month and stock value. Box plots display data consisting of values for one variable. Therefore, a box plot is not an appropriate choice. Both line plots and circle graphs are used to display frequencies within categorical data. Neither can be used for the given scenario. Scatter plots compare the values of two variables to see if there are any patterns present. Line graphs display two numerical variables on a coordinate grid and show trends among the variables.

		3	8	0

63. 380 minutes. To find the total driving time, the total distance of 350 miles can be divided by the constant speed of 60 miles per hour. This yields a time of 5.8333 hours, which is then rounded. Once the driving time is computed, the break times need to be found. If Gary takes a break for 10 minutes every 100 miles, he will take 3 breaks on his trip.

This will yield a total of 30 minutes of break time. Since the answer is needed in minutes, 5.8333 can be converted to minutes by multiplying by 60, giving a driving time of 350 minutes. Adding the break time of 30 minutes to the driving time of 350 minutes gives a total travel time of 380 minutes.

64. 66 Cookies. If the sixth graders bought 45% of the cookies, the number they bought is found by multiplying 0.45 by 500. They bought 225 cookies. The number of cookies left is:

$$500 - 225 = 275$$

During the second lunch, the seventh graders bought 40% of the cookies, which is found by multiplying 0.40 by the remaining 275 cookies. The seventh graders bought 110 cookies. This leaves 165 cookies to sell to the eighth graders. If they bought 60% of the remaining cookies, then they bought 99 cookies. Subtracting 99 from 165 cookies leaves Kelly with 66 cookies remaining after the three lunches.

65. 20. Because these are 2 parallel lines cut by a transversal, the angle with a measure of 45 degrees is equal to the measure of angle 6. Angle 6 and the angle labeled $5x + 35$ are supplementary to one another. The sum of these angles should be 180, so the following equation can be generated:

$$5x + 35 + 45 = 180$$

Solving for x, the sum of 35 and 45 is 80, which is then subtracted from 180 to yield 100. Dividing 100 by 5 gives the value of x, which is 20.

			1	0

66. 10. 8 squared is 64, and 6 squared is 36. These should be added together to get:

$$64 + 36 = 100$$

Then, the last step is to find the square root of 100 which is 10.

			2	4

67. 24; The long division would be completed as follows:

```
     24
 36|864
  - 72↓
    144
```

Answer Explanations for Practice Test #2

Editing/Revising

1. A: Choice *A* is the correct answer. The *'s* in *Juliet's* suggests a contraction using the word *is*. Proper nouns cannot be part of a contraction, so *Juliet's* should be changed to *Juliet is*. The word *well* is an adverb that modifies known, and *known* is an adjective. They should be joined with a hyphen to describe the word, *tragedy*.

2. F: Choice *F* is correct because sentence 2 is punctuated incorrectly. This sentence is a complex sentence, or a sentence consisting of a dependent and independent clause. The phrase, *When air is inhaled*, is a clause because it starts with the subordinating conjunction, *when*. The phrase cannot stand alone as a complete sentence, so a comma is needed after the word *inhaled*.

3. C: Choice *C* is correct. These two sentences aim to provide a visual of how far a driver can travel while reading a text. Combining these sentences clarifies the idea that the amount of time it takes to drive the length of a football field is the minimum amount of time it takes to read or send a text while driving.

4. H: Choice *H* is the most concise way to convey the idea presented in the sentence. Choice *E* is incorrect because the revision only eliminates some of the wordiness in the sentence. Choice *F* is incorrect because it eliminates too much information, such as Mr. Wilkinson's name and the fact that he was late to work. Choice *G* is also incorrect because it leaves out some important information, such as the fact that Mr. Wilkinson was late to work, and the class was being rewarded for good test grades.

5. A: Choice *A* is correct because the word *Flag* does not need to be capitalized, and commas are needed to separate three or more words in a list, such as *unity, strength, and courage*.

6. F: Choice *F* combines the two sentences while still including the correct information. Choice *E* leaves out the point that the density of a penguin's bones helps it to float and dive. Choice *G* alludes to the idea that birds just can't fly because they are fat. Choice *H* does not pertain to the information in either of the original two sentences.

7. C: Choice *C* is the correct answer. In sentence 3, the word *Libertys* is plural, indicating more than one Liberty. However, adding an apostrophe changes the word from a plural to a noun showing ownership, her crown. The word *weighs* is a singular verb in this sentence but needs to be a plural verb. Plural subjects should be followed by plural verbs and vice versa. So, since the subject of the sentence, *rays*, is plural, the verb *weighs* needs to be plural, or *weigh.*

8. G: Choice *G* correctly uses *from* to describe the fact that dogs are related to wolves. The word *through* is incorrectly used here, so Choice *E* is incorrect. Choice *F* makes no sense. Choice *H* unnecessarily changes the verb tense in addition to incorrectly using *through*.

9. B: Choice *B* is correct because the Oxford comma is applied, clearly separating the specific terms. Choice *A* lacks this clarity. Choice *C* is correct but too wordy since commas can be easily applied. Choice *D* doesn't flow with the sentence's structure.

10. H: Choice *H* correctly uses the question mark, fixing the sentence's main issue. Thus, Choice *E* is incorrect because questions do not end with periods. Choice *F*, although correctly written, changes the meaning of the original sentence. Choice *G* is incorrect because it completely changes the direction of the sentence, disrupts the flow of the paragraph, and lacks the crucial question mark.

11. A: Choice *A* is correct since there are no errors in the sentence. Choices *B* and *C* both have extraneous commas, disrupting the flow of the sentence. Choice *D* unnecessarily rearranges the sentence.

12. H: Choice *H* is correct because the commas serve to distinguish that *artificial selection* is just another term for *selective breeding* before the sentence continues. The structure is preserved, and the sentence can flow with more clarity. Choice *E* is incorrect because the sentence needs commas to avoid being a run-on. Choice *F* is close but still lacks the required comma after *selection*, so this is incorrect. Choice *G* is incorrect because the comma to set off the aside should be placed after *breeding* instead of *called*.

13. B: Choice *B* is correct because the sentence is talking about a continuing process. Therefore, the best modification is to add the word *to* in front of *increase*. Choice *A* is incorrect because this modifier is missing. Choice *C* is incorrect because, with the additional comma, the present tense of *increase* is inappropriate. Choice *D* makes more sense, but the tense is still not the best to use.

14. E: The sentence has no errors, so Choice *E* is correct. Choice *F* is incorrect because it adds an unnecessary comma. Choice *G* is incorrect because *advantage* should not be plural in this sentence without the removal of the singular *an*. Choice *H* is very tempting. While this would make the sentence more concise, this would ultimately alter the context of the sentence, which would be incorrect.

15. C: Choice *C* correctly uses *on to*, describing the way genes are passed generationally. The use of *into* is inappropriate for this context, which makes Choice *A* incorrect. Choice *B* is close, but *onto* refers to something being placed on a surface. Choice *D* doesn't make logical sense.

16. H: Choice *H* is correct, since only proper names should be capitalized. Because the name of a dog breed is not a proper name, Choice *E* is incorrect. In terms of punctuation, only one comma after *example* is needed, so Choices *F* and *G* are incorrect.

17. D: Choice *D* is the correct answer because "rather" acts as an interrupting word here and thus should be separated by commas. Choices *B* and *C* use commas unwisely, breaking the flow of the sentence.

18. F: Since the sentence can stand on its own without *Usually*, separating it from the rest of the sentence with a comma is correct. Choice *E* needs the comma after *Usually*, while Choice *G* uses commas incorrectly. Choice *H* is tempting but changing *turn* to past tense goes against the rest of the paragraph.

19. A: In Choice *A*, the dependent clause *Sometimes in particularly dull seminars* is seamlessly attached with a single comma after *seminars*. Choice *B* contains too many commas. Choice *C* does not correctly combine the dependent clause with the independent clause. Choice *D* introduces too many unnecessary commas.

20. H: Choice *H* rearranges the sentence to be more direct and straightforward, so it is correct. Choice *E* needs a comma after *on*. Choice *F* introduces unnecessary commas. Choice *G* creates an incomplete sentence, since *Because I wasn't invested in what was going on* is a dependent clause.

21. C: Choice *C* is fluid and direct, making it the best revision. Choice *A* is incorrect because the construction is awkward and lacks parallel structure. Choice *B* is incorrect because of the unnecessary comma and period. Choice *D* is close, but its sequence is still awkward and overly complicated.

22. F: Choice *F* correctly adds a comma after *person* and cuts out the extraneous writing, making the sentence more streamlined. Choice *E* is poorly constructed, lacking proper grammar to connect the sections of the sentence correctly. Choice *G* inserts an unnecessary semicolon and doesn't enable this section to flow well with the rest of the sentence. Choice *H* is better but still unnecessarily long.

23. D: This sentence, though short, is a complete sentence. The only thing the sentence needs is an em-dash after "Easy." In this sentence the em-dash works to add emphasis to the word "Easy" and also acts in place of a colon, but in a less formal way. Therefore, Choice *D* is correct. Choices *A* and *B* lack the crucial comma, while Choice *C* unnecessarily breaks the sentence apart.

24. G: Choice *G* successfully fixes the construction of the sentence, changing *drawing* into *to draw*. Keeping the original sentence disrupts the flow, so Choice *E* is incorrect. Choice *F*'s use of *which* offsets the whole sentence. Choice *H* is incorrect because it unnecessarily expands the sentence content and makes it more confusing.

25. B: Choice *B* fixes the homophone issue. Because the author is talking about people, *their* must be used instead of *there*. This revision also appropriately uses the Oxford comma, separating and distinguishing *lives, world, and future*. Choice *A* uses the wrong homophone and is missing commas. Choice *C* neglects to fix these problems and unnecessarily changes the tense of *applies*. Choice *D* fixes the homophone but fails to properly separate *world* and *future*.

Reading Comprehension

26. G: The poem uses imagery. William Carlos Williams is considered a Modernist poet who relied heavily on imagery to bring poetry to life. Poets like William Carlos Williams, Ezra Pound, and Marianne Moore wrote imagist poems typical of their time period. The poem is not metrical, but written in free verse, so Choice *E* is incorrect. Anaphora, Choice *F*, is a repetition of words at the beginning of a succession of lines. Synecdoche, Choice *H*, refers to a part of something that represents a whole.

27. B: This poem comes from the Modernist period. Modernism is a literary movement featuring writers like William Carlos Williams, Ezra Pound, T.S. Eliot, Marianne Moore, and Wallace Stevens. The poetry is a reaction to traditional metrical poetry and its outdated language. Modernists rely on heavy imagery and sometimes short, terse stanzas to create the impact of disjointedness with the self and language.

28. E: The poem acts as an extended metaphor of old age. Its juxtaposed imagery suggests the frailty of life ("Gaining and failing / they are buffeted / by a dark wind") alongside the fullness of life and its "piping of plenty." The other choices may have some aspects that are true of an effective analysis; however, the most important thing to recognize here is that the poem starts off very clearly as an extended metaphor.

29. D: The rhetorical device used in the last line is alliteration: "piping of plenty." Alliteration is where the same sound is used for an auditory effect. Metaphor and simile compare two things to each other, so these are incorrect. Anaphora is when two or more lines use the same beginning repetition.

30. G: The poem is describing winter. There is "dark wind," "bare trees," "snow," and a "snow glaze," so we can infer that this is the coldest season of the year.

31. B: The birds finally rest on weedstalks. We see this in the inverted line, "On harsh weedstalks / the flock has rested."

32. F: *E* is incorrect because it does not fit with the primary purpose of this passage, which is to tell a story of how a child plans to treat his parents when he sees the way they treat his grandfather. It is trying to remind readers to treat others with respect because that is how one wants to be treated, and that this does not apply only to elderly people. Choice *F* fits most appropriately with the primary purpose, since the son and wife see that they will be treated unfairly because they witness that their child plans to do it to them when they are older. To "reap what you sow" means that there are repercussions for every action. This may seem like the correct answer; however, the parents do not actually have to eat out of a trough later in life. They don't actually experience any repercussions. Even though it may be argued that the boy is being loyal to his grandfather, this does not fit with the primary purpose. The boy also never mentions that his actions are because he cares for his grandfather; rather, he simply mirrors the behaviors of his parents.

33. A: *A* is correct because it follows a series of events that happen in order, one right after the other. First the grandfather spills his food, then his son puts him in a corner, then the child makes a trough for his parents to eat out of when he's older, and finally the parents welcome the old man back to the table. Choice *B* is incorrect as even though it could be argued that the way they treat the old man is a problem, there really isn't a solution to the problem, even though they stop treating him badly. Also, problem and solution styles generally do not follow a chronological timeline. Choice *C* is incorrect because events in the passage are not compared and contrasted; this is not a primary organizational structure of the passage. Choice *D* is incorrect because there is no language to indicate that one person or event is more important than the other.

34. G: Although they do show him compassion in the end, it is not because they feel compassionate for him, but instead, it is because they recognize that their son plans to treat them the way they are treating the old man when they are older. So, they treat the old man the way they would want to be treated. Understanding is not the overall attitude they feel toward the old man, and it is only in realizing the cruelty of their behavior that they understand how they have been treating him. Choice *G* is correct because it condenses the actions of the son and his wife into a single word. Refusing to let the old man sit at the table when he clearly needs help and looks at the table with tear-filled eyes is a cruel thing to do. Choice *H* may be tempting to pick as they *are* impatient with him, but it's not the best answer. People can be impatient without being cruel.

35. C: Choice *A* is incorrect as there is no descriptive language to indicate that they are in the countryside. *B* is incorrect because the passage has no language or descriptions to indicate they are in America. Choice *C* is correct because the setting contains elements of a house: a table, a stove, and a corner. Choice *D* may be tempting as there is mention of "bits of wood upon the ground," but as there are no other elements of a forest in the story, this is not the correct answer.

36. H: They allow the old man to sit at the table because their son starts to make them a trough, so their motivation in letting him eat at the table is not because they feel sorry for him, but because they don't want their son to treat them that way when they are old. This makes Choice *E* incorrect. Their son did not tell them to let the old man sit at the table, so Choice *F* is incorrect. In the story, it mentions that even after the old man has eyes full of tears, the wife gave him a cheap wooden bowl to eat out of, so clearly his crying did not make them stop treating him badly, making Choice *G* incorrect. Choice *H* is correct because the parents let the old man sit at the table as a result of the boy mimicking their behavior.

37. B: Strong dislike. This vocabulary question can be answered using context clues. Based on the rest of the conversation, the reader can gather that Albert isn't looking forward to his marriage. As the Count notes that "you don't appear to me to be very enthusiastic on the subject of this marriage," and also remarks on Albert's "objection to a young lady who is both rich and beautiful," readers can guess Albert's feelings. The answer choice that most closely matches "objection" and "not . . . very enthusiastic" is *B*, "strong dislike."

38. G: Their name is more respected than the Danglars'. This inference question can be answered by eliminating incorrect answers. Choice *E* is tempting, considering that Albert mentions money as a concern in his marriage. However, although he may not be as rich as his fiancée, his father still has a stable income of 50,000 francs a year. Choice *F* isn't mentioned at all in the passage, so it's impossible to make an inference. Finally, Choice *H* is false because Albert's father arranged his marriage, but his mother doesn't approve of it. Evidence for Choice *G* can be found in the Count's comparison of Albert and Eugénie: "she will enrich you, and you will ennoble her." In other words, the Danglars are wealthier but the Morcef family has a more noble background.

39. D: Apprehensive. There are many clues in the passage that indicate Albert's attitude towards his marriage—far from enthusiastic, he has many reservations. This question requires test takers to understand the vocabulary in the answer choices. "Pragmatic" is closest in meaning to "realistic," and "indifferent" means "uninterested." The only word related to feeling worried, uncertain, or unfavorable about the future is "apprehensive."

40. F: He is like a wise uncle, giving practical advice to Albert. Choice *E* is incorrect because the Count's tone is friendly and conversational. Choice *G* is also incorrect because the Count questions why Albert doesn't want to marry a young, beautiful, and rich girl. While the Count asks many questions, he isn't particularly "probing" or "suspicious"—instead, he's asking to find out more about Albert's situation and then give him advice about marriage.

41. A: She belongs to a noble family. Though Albert's mother doesn't appear in the scene, there's more than enough information to answer this question. More than once is his family's noble background mentioned (not to mention that Albert's mother is the Comtess de Morcef, a noble title). The other answer choices can be eliminated—she is deeply concerned about her son's future; money isn't her highest priority because otherwise she would favor a marriage with the wealthy Danglars; and Albert describes her "clear and penetrating judgment," meaning she makes good decisions.

42. G: The richest people in society were also the most respected. The Danglars family is wealthier but the Morcef family has a more aristocratic name, which gives them a higher social standing. Evidence for the other answer choices can be found throughout the passage: Albert mentioned receiving money from his father's fortune after his marriage; Albert's father has arranged this marriage for him; and the Count speculates that Albert's mother disapproves of this marriage because Eugénie isn't from a noble background like the Morcef family, implying that she would prefer a match with a girl from aristocratic society.

43. A: He seems reluctant to marry Eugénie, despite her wealth and beauty. This is a reading comprehension question, and the answer can be found in the following lines: '"I confess," observed Monte Cristo, "that I have some difficulty in comprehending your objection to a young lady who is both rich and beautiful."' Choice *B* is the opposite (Albert's father is the one who insists on the marriage), Choice *C* incorrectly represents Albert's eagerness to marry, and Choice *D* describes a more positive attitude than Albert actually feels ("repugnance").

44. G: The author contrasts two different viewpoints, then builds a case showing preference for one over the other. Choice *E* is incorrect because the introduction does not contain an impartial definition, but rather, an opinion. Choice *F* is incorrect. There is no puzzling phenomenon given, as the author doesn't mention any peculiar cause or effect that is in question regarding poetry. Choice *H* does contain another's viewpoint at the beginning of the passage; however, to say that the author has no stake in this argument is incorrect; the author uses personal experiences to build their case.

45. B: Choice *B* accurately describes the author's argument in the text: that poetry is not irrelevant. While the author does praise, and even value, Buddy Wakefield as a poet, the author never heralds him as a genius. Eliminate Choice *A*, as it is an exaggeration. Not only is Choice *C* an exaggerated statement, but the author never mentions spoken word poetry in the text. Choice *D* is incorrect because this statement contradicts the writer's argument.

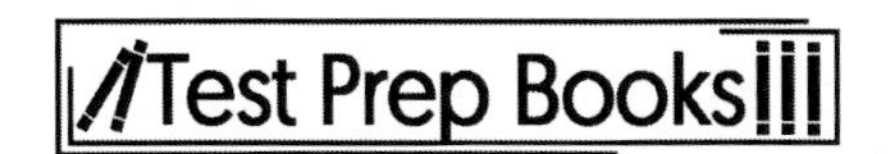

46. H: *Exiguously* means not occurring often, or occurring rarely, so Choice *H* would LEAST change the meaning of the sentence. Choice *E*, *indolently*, means unhurriedly, or slow, and does not fit the context of the sentence. Choice *F*, *inaudibly*, means quietly or silently. Choice *G*, *interminably*, means endlessly, or all the time, and is the opposite of the word *exiguously*.

47. D: A student's insistence that psychoanalysis is a subset of modern psychology is the most analogous option. The author of the passage tries to insist that performance poetry is a subset of modern poetry, and therefore, tries to prove that modern poetry is not "dying," but thriving on social media for the masses. Choice *A* is incorrect, as the author is not refusing any kind of validation. Choice *B* is incorrect; the author's insistence is that poetry will *not* lose popularity. Choice *C* mimics the topic but compares two different genres, while the author does no comparison in this passage.

48. F: The author's purpose is to disprove Gioia's article claiming that poetry is a dying art form that only survives in academic settings. In order to prove his argument, the author educates the reader about new developments in poetry (Choice *E*) and describes the brilliance of a specific modern poet (Choice *G*), but these serve as examples of a growing poetry trend that counters Gioia's argument. Choice *H* is incorrect because it contradicts the author's argument.

49. D: This question is difficult because the choices offer real reasons as to why the author includes the quote. However, the question specifically asks for the *main reason* for including the quote. The quote from a recently written poem shows that people are indeed writing, publishing, and performing poetry (Choice *B*). The quote also shows that people are still listening to poetry (Choice *C*). These things are true, and by their nature, serve to disprove Gioia's views (Choice *A*), which is the author's goal. However, Choice *D* is the most direct reason for including the quote, because the article analyzes the quote for its "complex themes" that "draws listeners and appreciation" right after it's given.

50. H: To enlighten the audience on the habits of sun-fish and their hatcheries. Choice *E* is incorrect because although the Adirondack region is mentioned in the text, there is no cause or effect relationships between the region and fish hatcheries depicted here. Choice *F* is incorrect because the text does not have an agenda, but rather is meant to inform the audience. Finally, Choice *G* is incorrect because the text says nothing of how sun-fish mate.

51. B: The word *wise* in this passage most closely means *manner*. Choices *A* and *C* are synonyms of *wise*; however, they are not relevant in the context of the text. Choice *D*, *ignorance*, is opposite of the word wise and is therefore incorrect.

52. E: Fish at the stage of development where they are capable of feeding themselves. Even if the word *fry* isn't immediately known to the reader, the context gives a hint when it says "until the fry are hatched out and are sufficiently large to take charge of themselves."

53. B: The sun-fish builds it with her tail and snout. The text explains this in the second paragraph: "she builds, with her tail and snout, a circular embankment 3 inches in height and 2 thick." Choice *A* is used in the text as a simile.

54. H: To conclude a sequence and add a final detail. The concluding sequence is expressed in the phrase "[t]he mother sun-fish, having now built or provided her 'hatchery.'" The final detail is the way in which the sun-fish guards the "inclosure." Choices *E, F,* and *G* are incorrect.

55. C: The text mentions all of the listed properties of minerals except the instance of minerals being organically formed. Objects or substances must be naturally occurring, must be a homogeneous solid, and must have a definite chemical composition in order to be considered a mineral.

56. E: Choice *E* is the correct answer because the prefix "homo" means same. Choice *F* is incorrect because "differing in some areas" would be linked to the root word "hetero," meaning "different" or "other."

57. C: Choice *C* is the correct answer because *-logy* refers to the study of a particular subject matter.

58. G: Choice *G* is the correct answer because the counterargument is necessary to point to the fact that researchers don't always agree with findings. Choices *E* and *F* are incorrect because the counterargument isn't overcomplicated or expressing bias, but simply stating an objective dispute. Choice *H* is incorrect because the counterargument is not used to persuade readers to create a new subsection of minerals.

59. A: Choice *D* can be eliminated because the Salem witch trials aren't even mentioned. While sympathetic to the plight of the accused, the author doesn't demand or urge the reader to demand reparations to the descendants; therefore, Choice *B* can also be ruled out. It's clear that the author's main goal is to educate the reader and shed light on the facts and hidden details behind the case. However, his focus isn't on the occult, but the specific Lancashire case itself. He goes into detail about suspects' histories and ties to Catholicism, revealing how the fears of the English people at the time sealed the fate of the accused witches. Choice *A* is correct.

60. F: It's important to note that these terms may not be an exact analog for *enduring*. However, through knowledge of the definition of *enduring*, as well as the context in which it's used, an appropriate synonym can be found. Plugging "circumstantial" into the passage in place of "enduring" doesn't make sense. Nor does "un-original" work, since this particular case of witchcraft stands out in history. "Wicked" is very descriptive, but this is an attribute applied to people, not events; therefore, this is an inappropriate choice as well. *Enduring* literally means long lasting, referring to the continued interest in this particular case of witchcraft. Therefore, it's a popular topic of 1600s witch trials, making "popular," Choice *F*, the best choice.

61. D: Choices A and B are irrelevant. The use of quotes lends credibility to the author. However, the presence of quotes alone doesn't necessarily mean that the author has a qualified perspective. What establishes the writer as a reliable voice is that the author's previous writing on the subject has been published before. This qualification greatly establishes the author's credentials as a historical writer, making Choice D the correct answer.

62. F: Choice *E* is incorrect, clearly taking the statement somewhat literally. The remaining three choices appear somewhat interconnected, and though they may be proven at some point later in the article, the focus must remain on the given excerpt. It's very possible that evidence was tampered with or even falsified, but this statement doesn't refer to this. While the author alludes that there may have been evidence tampering and potentially corruption, what the writer is directly saying is that the documentation of the court indicates an elaborate trial. It's clear that exaggerations may have taken place both during the case and in the written account. The reasoning behind this was to gain the attention of the people and even the crown. Choice *F* is the best answer because it not only aligns with the above statement, but ultimately encompasses the potentiality of Choices *G* and *H* as well.

63. C: Several of these answers could have contributed to the fear and political motivations around the Lancashire witch trials. What this answer's looking for is very specific: political motivations and issues that played a major role in the case. Choice C clearly outlines the public fears of the time. It also describes how the government can use this fear to weed out and eliminate traces of Catholicism (and witchcraft too). Catholicism and witchcraft were seen as dangerous and undermining to English Protestantism and governance. Choice D can be eliminated; while this information may have some truth and is certainly consistent with the general fear of witchcraft, the details about Lancashire's ancient history aren't mentioned in the text. Choice A is true but not necessarily political in nature. Choice B is very promising, though not outright mentioned.

64. H: The best evidence comes from Alizon herself. The text mentions that she confessed to bewitching John Law, thinking that she did him harm. From here she names her grandmother, who she believes corrupted her. Choice *F* can be ruled out; spectral evidence isn't mentioned. The case draws on knowledge of superstition of witchcraft, but this in itself can't be considered evidence, so Choice *E* is incorrect. Choice *G* isn't evidence in a modern sense; rumors have no weight in court and therefore are not evidence. While this is used as evidence to some degree, this still isn't the best evidence against Alizon and the witches.

65. D: Although Washington is from a wealthy background, the passage does not say that his wealth led to his republican ideals, so Choice *A* is not supported. Choice *B* also does not follow from the passage. Washington's warning against meddling in foreign affairs does not mean that he would oppose wars of every kind, so Choice *B* is incorrect. Choice *C* is also unjustified since the author does not indicate that Alexander Hamilton's assistance was absolutely necessary. Choice *D* is correct because the farewell address clearly opposes political parties and partisanship. The author then notes that presidential elections often hit a fever pitch of partisanship. Thus, it is follows that George Washington would not approve of modern political parties and their involvement in presidential elections.

66. E: The author finishes the passage by applying Washington's farewell address to modern politics, so the purpose probably includes this application. Choice *F* is incorrect because George Washington is already a well-established historical figure; furthermore, the passage does not seek to introduce him. Choice *G* is incorrect because the author is not fighting a common perception that Washington was merely a military hero. Choice *H* is incorrect because the author is not convincing readers. Persuasion does not correspond to the passage. Choice *E* states the primary purpose.

67. A: The tone in this passage is informative. Choice *B*, excited, is incorrect, because there are not many word choices used that would indicate excitement from the author. Choice *C*, bitter, is incorrect. Although the author does make a suggestion in the last paragraph to Americans, the statement is not necessarily bitter, but based on the preceding information. Choice *D*, comic, is incorrect, as the author does not try to make the audience laugh, nor do they make light of the situation in any way.

Math

1. D: To calculate the circumference of a circle, use the formula $2\pi r$, where r equals the radius, or half of the diameter, of the circle and $\pi = 3.14 \ldots$. Substitute the given information, $2\pi 5 = 31.4 \ldots$, which is Choice *D*.

2. E: To solve for x the steps are as follows:

$$4x - 12 = -2x$$

$$6x - 12 = 0$$

$$6x = 12$$

$$x = 2$$

3. B: There are two zeros for the given function. They are $x = 0\,, -2$. The zeros can be found a number of ways, but this particular equation can be factored into:

$$f(x) = x(x^2 + 4x + 4)$$

$$x(x + 2)(x + 2)$$

By setting each factor equal to zero and solving for x, there are two solutions. On a graph, these zeros can be seen where the line crosses the x-axis.

4. H: Dividing rational expressions follows the same rule as dividing fractions. The division is changed to multiplication by the reciprocal of the second fraction. This turns the expression into:

$$\frac{5x^3}{3x^2y} \times \frac{3y^9}{25}$$

Multiplying across and simplifying, the final expression is:

$$\frac{xy^8}{5}$$

5. B: The y-intercept of an equation is found where the x -value is zero. Plugging zero into the equation for x, the first two terms cancel out, leaving -4.

6. G: 216cm. Because area is a two-dimensional measurement, the dimensions are multiplied by a scale that is squared to determine the scale of the corresponding areas. The dimensions of the rectangle are multiplied by a scale of 3. Therefore, the area is multiplied by a scale of 3^2 (which is equal to 9):

$$24\ cm \times 9 = 216\ cm$$

7. C: 0.63

Divide 5 by 8, which results in 0.625. This rounds up to 0.63.

8. H: Finding the zeros for a function by factoring is done by setting the equation equal to zero, then completely factoring. Since there was a common x for each term in the provided equation, that would be factored out first. Then the quadratic that was left could be factored into two binomials, which are $(x + 1)(x - 4)$. Setting each factor equal to zero and solving for x yields three zeros.

9. B: This can be determined by finding the length and width of the shaded region. The length can be found using the length of the top rectangle, which is 18 inches, then subtracting the extra length of 4 inches and 1 inch. This means the length of the shaded region is 13 inches. Next, the width can be determined using the 6 inch measurement and subtracting the 2 inch measurement. This means that the width is 4 inches. Thus, the area is:

$$13 \times 4 = 52\, sq. in.$$

10. H: When an ordered pair is reflected over an axis, the sign of one of the coordinates must change. When it's reflected over the x-axis, the sign of the y-coordinate must change. The x-value remains the same. Therefore, the new ordered pair is $(-3, 4)$.

11. B: The equation can be solved by factoring the numerator into $(x + 6)(x - 5)$. Since that same factor $(x - 5)$ exists on top and bottom, that factor cancels. This leaves the equation $x + 6 = 11$. Solving the equation gives the answer $x = 5$. When this value is plugged into the equation, it yields a zero in the denominator of the fraction. Since this is undefined, there is no solution.

12. G: $\frac{19}{24}$

Set up the problem and find a common denominator for both fractions.

$$\frac{23}{24} - \frac{1}{6}$$

Multiply each fraction across by 1 to convert to a common denominator.

$$\frac{23}{24} \times \frac{1}{1} - \frac{1}{6} \times \frac{4}{4}$$

Once over the same denominator, subtract across the top.

$$\frac{23 - 4}{24} = \frac{19}{24}$$

13. C: Let *r* be the number of red cans and *b* be the number of blue cans. One equation is:

$$r + b = 10$$

The total price is $16, and the prices for each can means:

$$1r + 2b = 16$$

Multiplying the first equation on both sides by -1 results in:

$$-r - b = -10$$

Add this equation to the second equation, leaving $b = 6$. So, she bought 6 *blue* cans. From the first equation, this means $r = 4$; thus, she bought 4 *red* cans.

14. H: 270

Set up the division problem.

$$2.6\overline{)702}$$

Move the decimal over one place to the right in both numbers.

$$26\overline{)7020}$$

26 does not go into 7 but does go into 70 so start there.

$$\begin{array}{r} 270 \\ 26\overline{)7020} \\ \underline{-52} \\ 182 \\ \underline{-182} \\ 0 \end{array}$$

The result is 270

15. A: 2,504,774

Line up the numbers (the number with the most digits on top) to multiply. Begin with the right column on top and the right column on bottom.

Move one column left on top and multiply by the far right column on the bottom. Remember to add the carry over after you multiply. Continue that pattern for each of the numbers on the top row.

Starting on the far right column on top repeat this pattern for the next number left on the bottom. Write the answers below the first line of answers; remember to begin with a zero placeholder. Continue for each number in the top row.

Starting on the far-right column on top, repeat this pattern for the next number left on the bottom. Write the answers below the first line of answers. Remember to begin with zero placeholders.

Once completed, ensure the answer rows are lined up correctly, then add.

16. H: Recall the formula for area, area $=$ length $\times$ width. The answer must be in square inches, so all values must be converted to inches. Half of a foot is equal to 6 inches. Therefore, the area of the rectangle is equal to:

$$6 \text{ in} \times \frac{11}{2} \text{in} = \frac{66}{2} \text{in}^2 = 33 \text{ in}^2$$

17. B: The car is traveling at a speed of five meters per second. On the interval from one to three seconds, the position changes by ten meters. By making this change in position over time into a rate, the speed becomes ten meters in two seconds or five meters in one second.

18. E: $\frac{810}{2921}$

Line up the fractions.

$$\frac{15}{23} \times \frac{54}{127}$$

Multiply across the top and across the bottom.

$$\frac{15 \times 54}{23 \times 127} = \frac{810}{2921}$$

19. B: The exponent of the ten must be the same before any operations are performed on the numbers. So, $(2.6 \times 10^5) + (1.3 \times 10^4)$ cannot be added until one of the exponents on the ten is changed. The 1.3×10^4 can be changed to 0.13×10^5, then the 2.6 and 0.13 can be added. The answer is 2.73×10^5.

20. H: 3 times the sum of a number and 7 is greater than or equal to 32 can be translated into equation form utilizing mathematical operators and numbers.

21. C: To find the mean, or average, of a set of values, add the values together and then divide by the total number of values. Each day of the week has an adult ticket amount sold that must be added together. The equation is as follows:

$$\frac{22 + 16 + 24 + 19 + 29}{5} = 22$$

22. H: First, convert the distance that Courtney already drove to feet. Because there are three feet per yard, her distance traveled thus far in yards must be multiplied by 3:

$$1{,}236 \times 3 = 3{,}708 \text{ feet}$$

If the total distance to travel is 6,292 feet, there is $6292 - 3708 = 2{,}584$ feet left to travel.

23. A: The probability of choosing two customers simultaneously is the same as choosing one and then choosing a second without putting the first back into the pool of customers. This means that the probability of choosing a customer who bought cherry is $\frac{35}{100}$. Then without placing them back in the pool, it would be $\frac{34}{99}$.

So, the probability of choosing 2 customers simultaneously that both bought cherry would be:

$$\frac{35}{100} \times \frac{34}{99}$$

$$\frac{1{,}190}{9{,}900}$$

$$\frac{119}{990}$$

24. F: The number line shows:

$$x > -\frac{3}{4}$$

Each inequality must be solved for x to determine if it matches the number line. Choice *A* of $4x + 5 < 8$ results in $x < -\frac{3}{4}$, which is incorrect. Choice *C* of $-4x + 5 > 8$ yields $x < -\frac{3}{4}$, which is also incorrect. Choice *D* of $4x - 5 > 8$ results in $x > \frac{13}{4}$, which is not correct. Choice B, $-4x + 5 < 8$ is the only choice that results in the correct answer of:

$$x > -\frac{3}{4}$$

25. B: The perimeter of a rectangle is the sum of all four sides. Therefore, the answer is:

$$P = 14 + 8\frac{1}{2} + 14 + 8\frac{1}{2}$$

$$14 + 14 + 8 + \frac{1}{2} + 8 + \frac{1}{2} = 45 \text{ square inches.}$$

26. F: The least common multiple is the smallest number that is a multiple of two numbers. The first few multiples of 8 are 8, 16, 24, 32, 40, and 48. The first few multiples of 12 are 12, 24, 36, 48, and 60. Both 24 and 48 are common multiples of 8 and 12, but 24 is the least common multiple.

27. C: The equation used to find the slope of a line when given two points is as follows:

$$slope = \frac{y_2 - y_1}{x_2 - x_1}$$

Substituting the points into the equation yields:

$$\frac{8 - (-4)}{-5 - 10}$$

$$\frac{12}{-15}$$

$$-\frac{4}{5}$$

28. F: $12 \times 750 = 9{,}000$. Therefore, there are 9,000 milliliters of water, which must be converted to liters. 1,000 milliliters equals 1 liter; therefore, 9 liters of water are purchased.

29. B: First, subtract 9 from both sides to isolate the radical. Then, cube each side of the equation to obtain:

$$2x + 11 = 27$$

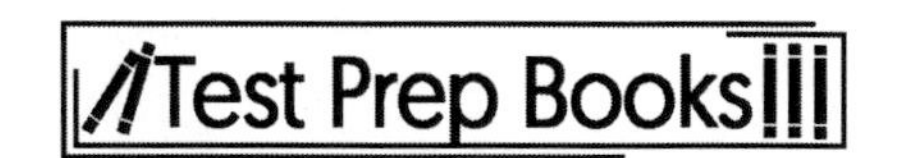

Subtract 11 from both sides, and then divide by 2. The result is $x = 8$. Plug 8 back into the original equation to obtain the true statement to check the answer:

$$\sqrt[3]{16 + 11} + 9 = 12$$

$$\sqrt[3]{27} + 9 = 12$$

$$3 + 9 = 12$$

30. E: Operations within the parentheses must be completed first. Then, division is completed. Finally, addition is the last operation to complete. When adding decimals, digits within each place value are added together. Therefore, the expression is evaluated as:

$$(2 \times 20) \div (7 + 1) + (6 \times 0.01) + (4 \times 0.001)$$

$$40 \div 8 + 0.06 + 0.004$$

$$5 + 0.06 + 0.004 = 5.064$$

31. B: The system can be solved using substitution. Solve the second equation for *y*, resulting in:

$$y = 1 - 2x$$

Plugging this into the first equation results in the quadratic equation:

$$x^2 - 2x + 1 = 4$$

In standard form, this equation is equivalent to $x^2 - 2x - 3 = 0$ and in factored form is:

$$(x - 3)(x + 1) = 0$$

Its solutions are $x = 3$ and $x = -1$. Plugging these values into the second equation results in $y = -5$ and $y = 3$, respectively. Therefore, the solutions are the ordered pairs $(-1, 3)$ and $(3, -5)$.

32. F: Because the 65-degree angle and angle *b* sum to 180 degrees, the measurement of angle *b* is 115 degrees. Because of corresponding angles, angle *b* is equal to angle *f*. Therefore, angle *f* measures 115 degrees.

33. C: Finding the product means distributing one polynomial to the other so that each term in the first is multiplied by each term in the second. Then, like terms can be collected. Multiplying the factors yields the expression:

$$x^3 + 5x^2 - 6x + 2x^2 + 10x - 12$$

Collecting like terms means adding the x^2 terms and adding the x terms. The final answer after simplifying the expression is:

$$x^3 + 7x^2 + 4x - 12$$

34. G: A dollar contains 20 nickels. Therefore, if there are 12 dollars' worth of nickels, there are:

$$12 \times 20 = 240 \text{ nickels}$$

Each nickel weighs 5 grams. Therefore, the weight of the nickels is:

$$240 \times 5 = 1{,}200 \text{ grams}$$

Adding in the weight of the empty piggy bank, the filled bank weighs 2,250 grams.

35. B: This problem can be solved using the Pythagorean Theorem. The triangle has a hypotenuse of 15 and one leg of 12. These values can be substituted into the Pythagorean formula to yield:

$$12^2 + b^2 = 15^2$$

$$144 + b^2 = 225$$

$$81 = b^2$$

$$b = 9$$

In this problem, b is represented by x so $= 9$ is the correct answer.

36. H: 3 must be multiplied times $27\frac{3}{4}$. In order to easily do this, the mixed number should be converted into an improper fraction.

$$27\frac{3}{4} = \frac{27 \times 4 + 3}{4} = \frac{111}{4}$$

Therefore, Denver had approximately

$$\frac{3 \times 111}{4} = \frac{333}{4} \text{ inches of snow}$$

The improper fraction can be converted back into a mixed number through division:

$$\frac{333}{4} = 83\frac{1}{4} \text{ inches}$$

37. B: The function presented is being evaluated for $x + 1$; therefore, $x + 1$ must be substituted into the original function as follows:

$$f(x + 1) = (x + 1)^2 - 3(x + 1) + 17$$

The squared portion of the function becomes $x^2 + 2x + 1$, and distributing the -3 results in:

$$f(x + 1) = x^2 + 2x + 1 - 3x - 3 + 17$$

Combining like terms results in:

$$x^2 - x + 15$$

38. F: When the number of data points provided is an even number, then the average of the two middle points is the median. Each set of responses provided should be ordered from least to greatest, and then the middle two values should be averaged together to see which set provides a median of 14. Choice *F*, when ordered, is 11, 12, 13, 15, 16, and 16. The middle two values averaged together is $\frac{13+15}{2} = 14$ miles, which is the correct answer.

39. C: These two events are mutually exclusive because the students only picked one flavor of ice cream as the favorite so a student can't have chosen two flavors. Therefore, the probability of choosing a student who likes each flavor of interest (vanilla and strawberry) should be added together. 30% of students chose vanilla, and 20% of the students chose strawberry. Expressed as percentages the probabilities are $\frac{3}{10}$ and $\frac{2}{10}$ which can be added together to find:

$$\frac{3}{10} + \frac{2}{10} = \frac{5}{10} = \frac{1}{2}$$

40. F: Let x represent the price of the stereo system. Complete the calculation as follows:

$$12\ percent = 0.12$$
$$156 = 0.12x$$
$$x = \$1{,}300$$

41. A: 13 nurses

Using the given information of 1 nurse to 25 patients and 325 patients, set up an equation to solve for number of nurses (N):

$$\frac{N}{325} = \frac{1}{25}$$

Multiply both sides by 325 to get N by itself on one side.

$$\frac{N}{1} = \frac{325}{25} = 13\ nurses$$

42. E: 12

Calculate how many gallons the bucket holds.

$$11.4\ L \times \frac{1\ gal}{3.8\ L} = 3\ gal$$

Now how many buckets to fill the pool which needs 35 gallons.

$$35 \div 3 = 11.67$$

Since the amount is more than 11 but less than 12, we must fill the bucket 12 times.

43. C: $x = 150$

Set up the initial equation.

$$\frac{2x}{5} - 1 = 59$$

Add 1 to both sides.

$$\frac{2x}{5} - 1 + 1 = 59 + 1$$

Multiply both sides by 5/2.

$$\frac{2x}{5} \times \frac{5}{2} = 60 \times \frac{5}{2} = 150$$

$$x = 150$$

44. G: $51.93

List the givens.

$$Tax = 6.0\% = 0.06$$

$$Sale = 50\% = 0.5$$

$$Hat = \$32.99$$

$$Jersey = \$64.99$$

Calculate the sales prices.

$$Hat\ Sale = 0.5\,(32.99) = 16.495$$

$$Jersey\ Sale = 0.5\,(64.99) = 32.495$$

Total the sales prices.

$$Hat\ sale + jersey\ sale = 16.495 + 32.495 = 48.99$$

Calculate the tax and add it to the total sales prices.

$$Total\ after\ tax = 48.99 + (48.99\ x\ 0.06) = \$51.93$$

45. D: $0.45

List the givens.

$$Store\ coffee = \$1.23/lbs$$

$$Local\ roaster\ coffee = \$1.98/1.5\ lbs$$

Calculate the cost for 5 lbs of store brand.

$$\frac{\$1.23}{1\ lbs} \times 5\ lbs = \$6.15$$

Calculate the cost for 5 lbs of the local roaster.

$$\frac{\$1.98}{1.5\ lbs} \times 5\ lbs = \$6.60$$

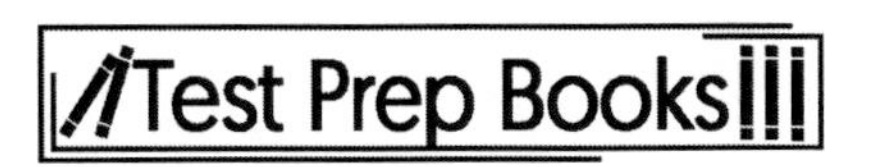

Subtract to find the difference in price for 5 lbs.

$$\begin{array}{r} \$6.60 \\ -\$6.15 \\ \hline \$0.45 \end{array}$$

46. H: $3,325

List the givens.

$$1{,}800\ ft. = \ \$2{,}000$$

$$Cost\ after\ 1{,}800\ ft. = \ \$1.00/ft.$$

Find how many feet left after the first 1,800 ft.

$$\begin{array}{r} 3{,}125\text{ ft.} \\ -\ 1{,}800\text{ ft.} \\ \hline 1{,}325\text{ ft.} \end{array}$$

Calculate the cost for the feet over 1,800 ft.

$$1{,}325\ ft.\times\frac{\$1.00}{1\ ft} = \$1{,}325$$

Add these together to find the total for the entire cost.

$$\$2{,}000\ +\ \$1{,}325\ =\ \$3{,}325$$

47. D: 290 beds

Using the given information of 2 beds to 1 room and 145 rooms, set up an equation to solve for number of beds (B):

$$\frac{B}{145} = \frac{2}{1}$$

Multiply both sides by 145 to get B by itself on one side.

$$\frac{B}{1} = \frac{290}{1} = 290\ beds$$

48. H: The length of LM can be found by a series of calculations:

$$KL + LM = 16 \qquad KL = 16 - LM$$

$$LM + MN = 20 \qquad MN = 20 - LM$$

$$KN = KL + MN + LM = 30$$

$$16 - LM + 20 - LM + LM = 30$$

$$36 - 30 = LM$$

$$6 = LM$$

49. B: The three angles lie on a straight line; therefore, the sum of all the angles must equal $180°$. The values for angle x and angle y should be added together and subtracted from $180°$ to find the value for angle z as follows:

$$180 - (48° + 2(48°)) = 36°$$

50. E: The first step is to evaluate the exponent inside the absolute value symbols. $(-5)^3$ yields -125. The next step is to evaluate the terms inside each of the two sets of absolute value symbols: $|-5.6| + |137.3|$. The absolute value of -5.6 is 5.6 and the absolute value of $|137.3|$ is 137.3 so the answer is:

$$5.6 + 137.3 = 142.9$$

51. D: The values for x and y should be plugged into the equation to find the correct answer.

$$3.2(2.6)(5.3) - 4.1(5.3) = 22.366$$

52. G: Each value can be calculated so that they can be compared to find which one is the greatest. The mean is equal to:

$$\frac{26 + 27 + 27 + 29 + 30 + 32 + 33 + 33 + 33 + 35}{10} = 30.5$$

The median is equal to:

$$\frac{30 + 32}{2} = 31$$

The mode is equal to 33 because that number occurs 3 times in the data set. The range is equal to:

$$35 - 26 = 9$$

Therefore, the mode is the greatest value of the answer choices.

53. B: This inequality can be seen with the use of a number line. $\frac{3}{7}$ is close to $\frac{1}{2}$.

$\frac{5}{6}$ is close to 1, but less than 1, and $\frac{8}{7}$ is greater than 1. Therefore, $\frac{3}{7}$ is less than $\frac{5}{6}$.

54. G: Each number in the sequence is adding one more than the difference between the previous two.

For example, $10 - 6 = 4, 4 + 1 = 5$.

Therefore, the next number after 10 is $10 + 5 = 15$.

Going forward, $21 - 15 = 6, 6 + 1 = 7$. The next number is $21 + 7 = 28$.

Therefore, the difference between numbers is the set of whole numbers starting at 2: 2, 3, 4, 5, 6, 7....

55. B: This is a division problem because the original amount needs to be split up into equal amounts. The mixed number $11\frac{1}{2}$ should be converted to an improper fraction first:

$$11\frac{1}{2} = \frac{(11 \times 2) + 1}{2} = \frac{23}{2}$$

Carey needs determine how many times $\frac{23}{2}$ goes into 184. This is a division problem:

$$184 \div \frac{23}{2} = ?$$

The fraction can be flipped, and the problem turns into the multiplication:

$$184 \times \frac{2}{23} = \frac{368}{23}$$

This improper fraction can be simplified into 16 because $368 \div 23 = 16$. The answer is 16 lawn segments.

56. E: The additive and subtractive identity is 0. When added or subtracted to any number, 0 does not change the original number.

57. B: Each hour on the clock represents 30 degrees. For example, 3:00 represents a right angle. Therefore, 5:00 represents 150 degrees.

58. H: $27\frac{7}{22}$

Set up the division problem.

$$44 \overline{\smash{)}\,1202}$$

44 does not go into 1 or 12 but will go into 120 so start there.

			2	7
44	1	2	0	2
	-	8	8	
		3	2	2
	-	3	0	8
			1	4

The answer is $27\frac{14}{44}$.

Reduce the fraction for the final answer.

$$27\frac{7}{22}$$

59. A: Using the order of operations, multiplication and division are computed first from left to right. Multiplication is on the left; therefore, multiplication should be performed first.

60. H: A factor of 36 is any number that can be divided into 36 and have no remainder. $36 = 36 \times 1, 18 \times 2, 9 \times 4,$ and 6×6. Therefore, it has 7 unique factors: 36, 18, 9, 6, 4, 2, and 1.

61. A: Compare each numeral after the decimal point to figure out which overall number is greatest. In answers *A* (1.43785) and *C* (1.43592), both have the same tenths (4) and hundredths (3). However, the thousandths is greater in answer *A* (7), so *A* has the greatest value overall.

62. F: Using the conversion rate, multiply the projected weight loss of 25 lb. by 0.45 $\frac{kg}{lb}$ to get the amount in kilograms (11.25 kg).

			1	8

63. 18; If Ray will be 25 in three years, then he is currently 22. The problem states that Lisa is 13 years younger than Ray, so she must be 9. Sam's age is twice that, which means that the correct answer is 18.

	5	0	.	7

64. 50.7; The values for the missing sides must first be found before the perimeter can be calculated. The missing side that is the hypotenuse of the right triangle can be calculated using the Pythagorean Theorem as follows:

$$11^2 + 4^2 = x^2$$

$$121 + 16 = x^2$$

$$137 = x^2$$

$$x = 11.7$$

The other missing side is equal to the value of the length of the larger rectangle less than the value of the side of the square $12 - 4 = 8$. Then, all the sides can be added together to find the perimeter:

$$12 + 6 + 8 + 5 + 4 + 4 + 11.7 = 50.7$$

			1	5

65. 15; Follow the *order of operations* in order to solve this problem. Solve the parentheses first, and then follow the remainder as usual.

$$(6 \times 4) - 9$$

This equals $24 - 9$ or 15.

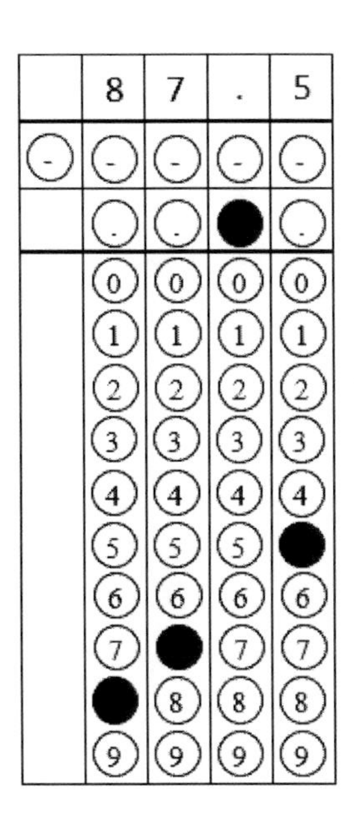

66. 87.5; For an even number of total values, the *median* is calculated by finding the *mean* or average of the two middle values once all values have been arranged in ascending order from least to greatest. In this case, $(92 + 83) \div 2$ would equal the median 87.5.

				6
-	-	-	-	-
	.	.	.	.
	0	0	0	0
	1	1	1	1
	2	2	2	2
	3	3	3	3
	4	4	4	4
	5	5	5	5
	6	6	6	●
	7	7	7	7
	8	8	8	8
	9	9	9	9

67. 6; The formula for the perimeter of a rectangle is $P = 2L + 2W$, where P is the perimeter, L is the length, and W is the width. The first step is to substitute all of the data into the formula:

$$36 = 2(12) + 2W$$

Simplify by multiplying 2×12:

$$36 = 24 + 2W$$

Simplifying this further by subtracting 24 on each side, which gives:

$$36 - 24 = 24 - 24 + 2W$$

$$12 = 2W$$

Divide by 2:

$$6 = W$$

The width is 6 cm. Remember to test this answer by substituting this value into the original formula:

$$36 = 2(12) + 2(6)$$

SHSAT Practice Test #3

Editing/Revising

Editing/Revising Part A

1. Read this paragraph.

    (1) The word *tsunami* means "killer wave" in Japanese. (2) Tsunamis are a series of giant waves that form in the ocean, and begin traveling to the shore at hundreds of km/h. (3) Tsunamis can be up to 100 feet high by the time they reach the shore. (4) Tsunamis hit the shore with such speed and force that the destruction affects miles and miles of land beyond the shore.

Which sentence should be revised to correct an error in sentence structure?

a. Sentence 1
b. Sentence 2
c. Sentence 3
d. Sentence 4

2. Read this paragraph.

    (1) Susan Eloise Hinton was seventeen when The Outsiders was published, in 1967. (2) The novel is inspired by two rival gangs Hinton observed at her high school. (3) Because she was concerned that male readers would not read a book about men in gangs written by a woman, Hinton opted to publish the book using the pen name S.E. Hinton. (4) Hinton's book has since sold over 14 million copies and has been released as a movie.

How should the paragraph be revised?

e. Sentence 1: Change The Outsiders to *The Outsiders* AND remove the comma after ***published***.
f. Sentence 2: Change ***observed*** to **observes** AND ***high school*** to **High School**.
g. Sentence 3: Remove the comma after ***woman*** AND change ***pen*** to **pin**.
h. Sentence 4: Change ***Hinton's*** to **Hintons** AND ***has been*** to **was**.

3. Read this paragraph.

    (1) Did you know that it literally pays to recycle? (2) An estimated 36 billion aluminum cans were disposed of in landfills last year. (3) Recycling this many cans would have earned someone almost $600 million. (4) While collecting such a large number of cans would be impossible for one person collecting them around your city might be an idea worth entertaining.

Which sentence should be revised to correct an error in sentence structure?

a. Sentence 1
b. Sentence 2
c. Sentence 3
d. Sentence 4

4. Read these sentences.

Back-to-school shopping can be very expensive, especially for those buying for multiple children.

Texas offers a tax-free weekend two weeks before school starts.

What is the best way to combine the sentences to clarify the relationship between the ideas?

e. Texas offers a tax-free weekend for parents doing back-to-school shopping for multiple children.
f. For parents who are buying school clothes for multiple children, Texas offers a tax-free weekend two weeks before school starts.
g. Two weeks before school starts, parents can do school shopping for multiple children during tax-free weekend.
h. Since back-to-school shopping can be very expensive, Texas hosts a statewide tax-free weekend to help ease the cost of buying children's clothes and school supplies.

5. Read this paragraph.

(1) One of the most well-known and most watched video platforms is YouTube. (2) In February 2005, the reality of YouTube comes to life, and the first video is published just two months later. (3) Two years later, Google purchased YouTube for $1.65 billion. (4) Today, there are 98 versions of YouTube, with at least 80% of Americans watching at least one video each month.

How should the paragraph be revised?

a. Sentence 1: Change ***well-known*** to **well known** AND ***platforms*** to **platform**.
b. Sentence 2: Change ***comes*** to **came** AND ***is*** to **was**.
c. Sentence 3: Change ***$1.65*** to **$1,65** AND ***billion*** to **billions**.
d. Sentence 4: Remove the comma after ***Today*** AND change ***%*** to **percent**.

6. Read these sentences.

Chocolate has a melting point of 93°F.

The average body temperature is 97°F.

Chocolate melts in your mouth.

What is the best way to combine the sentences to clarify the relationship between the ideas?

e. Your mouth is almost the same temperature as chocolate, so it doesn't melt easily in your mouth.
f. Chocolate melts in your mouth because your body temperature is only 97°F.
g. Chocolate melts easily in your mouth because it has a melting point lower than the average body temperature.
h. If the melting point of chocolate was 4° higher, it would not melt very well in your mouth.

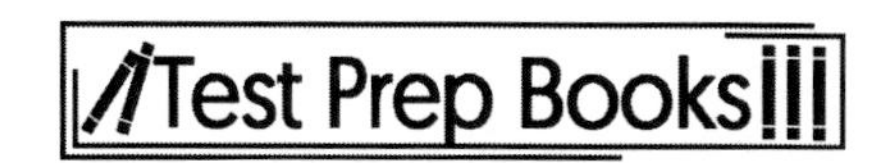

7. Read this paragraph.

(1) Hurricanes are unpredictable and potentially dangerous storms. (2) When there is a storm surge of the ocean, the result is a hurricane producing flooding and high winds. (3) Hurricanes have been known to cause extensive damage, not just to individual homes but to entire cities. (4) In some instance, hurricanes can cause serious bodily injurys or even death.

How should the paragraph be revised?

a. Sentence 1: Change ***potentially*** to **potential** AND ***storms*** to **storm**.
b. Sentence 2: Change ***of*** to **in** AND ***result*** to **results**.
c. Sentence 3: Change ***extensive*** to **extinsive** AND ***damage*** to **damages**.
d. Sentence 4: Change ***instance*** to **instances** AND ***injurys*** to **injuries**.

8. Read this paragraph.

(1) Giraffes live in the sub-Saharan region of Africa, primarily in the savanna areas. (2) Their height allows them to eat from trees that are much higher than other animals can reach. (3) Giraffes have long tongues to help them pull leaves from branches. (4) Because giraffes eat over 100 pounds of twigs and leaves a day their tongues are exposed to the sun a great deal of the time, so their tongues are black to keep them from getting sunburned.

Which sentence should be revised to correct an error in sentence structure?

e. Sentence 1
f. Sentence 2
g. Sentence 3
h. Sentence 4

Editing/Revising Part B

Questions 9–17 are based on the following passage:

The name "Thor" has always been associated with great power. (1) Arguably, Norse Mythologies most popular and powerful god is Thor of the Aesir. My first experience of Thor was not like most of today's generation. I grew up reading Norse mythology where (2) Thor wasn't a comic book superhero, but even mightier. There are stories of Thor destroying mountains, (3) defeating scores of giants and lifting up the world's largest creature the Midgard Serpent. But always, Thor was a protector.

Like in modern comics and movies, Thor was the god of thunder and wielded (4) the hammer Mjolnir however there are several differences between the ancient legend and modern hero. (5) For example, Loki, the god of mischief, isn't Thor's brother. Loki is actually Thor's servant, but this doesn't stop the trickster from causing chaos, chaos that Thor has to then quell. In all of his incarnations, Thor is a god that reestablishes order by tempering the chaos around him. (6) This is also symbolized in his prized weapon Mjolnir a magic hammer. A hammer is both a weapon and a (7) tool, but why would a god favor a seemingly everyday object?

A hammer is used to shape metal and create change. The hammer tempers raw iron, (8) ore that is in an chaotic state of impurities and shapelessness, to create an item of worth. Thus, a hammer is in many ways a tool that brings a kind of order to the world—like Thor. Hammers were also tools of everyday people, which further endeared Thor to the common man. Therefore, it's no surprise that Thor remains an iconic hero to this day.

I began thinking to myself, why is Thor so prominent in our culture today even though many people don't follow the old religion? (9) Well the truth is that every culture throughout time, including ours, needs heroes. People need figures in their lives that give them hope and make them aspire to be great. We need the peace of mind that chaos will eventually be brought to order and that good can conquer evil. Thor was a figure of hope and remains so to this day.

9. Which of the following would be the best choice for this sentence (reproduced below)?

(1) Arguably, Norse Mythologies most popular and powerful god is Thor of the Aesir.

a. NO CHANGE
b. Arguably Norse Mythologies most
c. Arguably, Norse mythology's most
d. Arguably, Norse Mythology's most

10. Which of the following would be the best choice for this sentence (reproduced below)?

I grew up reading Norse mythology where (2) Thor wasn't a comic book superhero, but even mightier.

e. NO CHANGE
f. Thor wasn't a comic book superhero. He was even mightier.
g. Thor wasn't a comic book superhero but even mightier.
h. Thor wasn't a comic book superhero, he was even mightier.

11. Which of the following would be the best choice for this sentence (reproduced below)?

There are stories of Thor destroying mountains, (3) defeating scores of giants and lifting up the world's largest creature the Midgard Serpent.

a. NO CHANGE
b. defeating scores of giants, and lifting up the world's largest creature, the Midgard Serpent.
c. defeating scores of giants, and lifting up the world's largest creature the Midgard Serpent.
d. defeating scores, of giants, and lifting up the world's largest creature the Midgard Serpent.

12. Which of the following would be the best choice for this sentence (reproduced below)?

Like in modern comics and movies, Thor was the god of thunder and wielded (4) the hammer Mjolnir however there are several differences between the ancient legend and modern hero.

e. NO CHANGE
f. the hammer Mjolnir, however there are several differences
g. the hammer Mjolnir. However there are several differences
h. the hammer Mjolnir. However, there are several differences

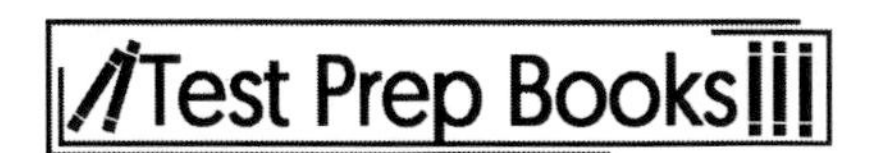

13. Which of the following would be the best choice for this sentence (reproduced below)?

(5) For example, Loki, the god of mischief, isn't Thor's brother.

a. NO CHANGE
b. For example, Loki the god of mischief isn't Thor's brother.
c. For example, Loki the god of mischief, isn't Thor's brother.
d. For example Loki, the god of mischief, isn't Thor's brother.

14. Which of the following would be the best choice for this sentence (reproduced below)?

(6) This is also symbolized in his prized weapon Mjolnir a magic hammer.

e. NO CHANGE
f. This is also symbolized in his prized weapon, Mjolnir a magic hammer.
g. This is also symbolized in his prized weapon, Mjolnir, a magic hammer.
h. This is also symbolized in his prized weapon Mjolnir, a magic hammer.

15. Which of the following would be the best choice for this sentence (reproduced below)?

A hammer is both a weapon and a (7) tool, but why would a god favor a seemingly everyday object?

a. NO CHANGE
b. tool; why would a god favor a seemingly everyday object?
c. tool, but, why would a god favor a seemingly everyday object?
d. tool, however, why would a god favor a seemingly everyday object?

16. Which of the following would be the best choice for this sentence (reproduced below)?

The hammer tempers raw iron, (8) ore that is in an chaotic state of impurities and shapelessness, to create an item of worth.

e. NO CHANGE
f. ore that is in a chaotic state of impurities and shapelessness
g. ore that has the impurities and shapelessness of a chaotic state
h. ore that is in an chaotic state, of impurities and shapelessness,

17. Which of the following would be the best choice for this sentence (reproduced below)?

(9) Well the truth is that every culture throughout time, including ours, needs heroes.

a. NO CHANGE
b. Well, the truth is, every culture throughout time, including ours,
c. Well, every culture throughout time, including ours, in truth
d. Well, the truth is that every culture throughout time, including ours,

Questions 18–25 are based on the following passage:

In our essay and class discussion, (10) we came to talking about mirrors. It was an excellent class in which we focused on an article written by Salman Rushdie that compared the homeland to a mirror. (11) Essentially this mirror was an metaphor for us and our homeland. (12) When we look at our reflection we see the culture, our homeland staring back at us. An interesting analogy, but the conversation really began when we read that Rushdie himself stated that the cracked mirror is more valuable than a whole one. But why?

(13) After reflecting on the passage I found the answer to be simple. The analogy reflects the inherent nature of human individuality. The cracks in the mirror represent different aspects of our own being. Perhaps it is our personal views, our hobbies, or our differences with other people, but (14) whatever it is that makes us unique defines us, even while we are part of a big culture. (15) What this tells us is that we can have a homeland, but ultimately we ourselves are each different in it.

Just because one's (16) mirror is cracked, the individuals isn't disowned from the actual, physical homeland and culture within. It means that the homeland is uniquely perceived by the (17) individual beholding it and that there are in fact many aspects to culture itself. Like the various cracks, a culture has religion, language, and many other factors that form to make it whole. What this idea does is invite the viewer to accept their own view of their culture as a whole.

Like in Chandra's *Love and Longing in Bombay*, a single homeland has many stories to tell. Whether one is a cop or a retired war veteran, the individual will perceive the different aspects of the world with unduplicated eyes. (18) Rushdie, seems to be urging his readers to love their culture but to not be pressured by the common crowd. Again, the cracks represent differences which could easily be interpreted as views about the culture, so what this is saying is to accept the culture but accept oneself as well.

From the essay "Portals to Homeland: Mirrors"

18. Which of the following would be the best choice for this sentence (reproduced below)?

In our essay and class discussion, (10) we came to talking about mirrors.

e. NO CHANGE
f. we were talking about
g. we talked about
h. we came to talk about

19. Which of the following would be the best choice for this sentence (reproduced below)?

(11) Essentially this mirror is an metaphor for us and our homeland.

a. NO CHANGE
b. Essentially, this mirror is a metaphor for us and our homeland.
c. Essentially, this mirror is an metaphor for us and our homeland.
d. Essentially this mirror is an metaphor, for us and our homeland.

20. Which of the following would be the best choice for this sentence (reproduced below)?

(12) When we look at our reflection we see the culture, our homeland staring back at us.

e. NO CHANGE
f. When we look at our reflection we see our culture our homeland staring back at us.
g. When we look at our reflection we saw our culture, our homeland, staring back at us.
h. Looking at our reflection we see our culture as our homeland is staring back at us.

21. Which of the following would be the best choice for this sentence (reproduced below)?

(13) After reflecting on the passage I found the answer to be simple.

a. NO CHANGE
b. After reflecting on the passage; I found the answer to be simple.
c. After reflecting on the passage I finding the answer to be simple.
d. After reflecting on the passage, I found the answer to be simple.

22. Which of the following would be the best choice for this sentence (reproduced below)?

Perhaps it is our personal views, our hobbies, or our differences with other people, but (14) whatever it is that makes us unique defines us, even while we are part of a big culture.

e. NO CHANGE
f. whatever it is, that makes us unique, defines us, even while we are part of a big culture.
g. whatever it is that makes us unique also defines us, even while we are part of a bigger culture.
h. whatever it is that makes us unique defines us, even though we are part of a big culture.

23. Which of the following would be the best choice for this sentence (reproduced below)?

(15) What this tells us is that we can have a homeland, but ultimately we ourselves are each different in it.

a. NO CHANGE
b. What this tells us is that we can have a homeland, but ultimately, we ourselves are each different in it.
c. What this tells us is that we can have a homeland, however, ultimately, we ourselves are each different in it.
d. What this tells us is that we can have a homeland, ultimately we ourselves are each different in it.

24. Which of the following would be the best choice for this sentence (reproduced below)?

Just because one's (16) mirror is cracked, the individuals isn't disowned from the actual, physical homeland and culture within.

e. NO CHANGE
f. mirror is cracked, the individuals will not be disowned
g. mirror is cracked, the individuals aren't disowned
h. mirror is cracked, the individual isn't disowned

25. Which of the following would be the best choice for this sentence (reproduced below)?

It means that the homeland is uniquely perceived by the (17) individual beholding it and that there are, in fact, many aspects of culture itself.

a. NO CHANGE
b. individual beholding it; and that there are in fact many aspects
c. individual beholding it and that there is, in fact, many aspects
d. individual beholding it and there's in fact, many aspects

Reading Comprehension

Questions 26–31 are based on the following passage:

As long ago as 1860 it was the proper thing to be born at home. At present, so I am told, the high gods of medicine have decreed that the first cries of the young shall be uttered upon the anesthetic air of a hospital, preferably a fashionable one. So young Mr. and Mrs. Roger Button were fifty years ahead of style when they decided, one day in the summer of 1860, that their first baby should be born in a hospital. Whether this anachronism had any bearing upon the astonishing history I am about to set down will never be known.

I shall tell you what occurred, and let you judge for yourself.

The Roger Buttons held an enviable position, both social and financial, in ante-bellum Baltimore. They were related to the This Family and the That Family, which, as every Southerner knew, entitled them to membership in that enormous peerage which largely populated the Confederacy. This was their first experience with the charming old custom of having babies—Mr. Button was naturally nervous. He hoped it would be a boy so that he could be sent to Yale College in Connecticut, at which institution Mr. Button himself had been known for four years by the somewhat obvious nickname of "Cuff."

On the September morning consecrated to the enormous event he arose nervously at six o'clock dressed himself, adjusted an impeccable stock, and hurried forth through the streets of Baltimore to the hospital, to determine whether the darkness of the night had borne in new life upon its bosom.

When he was approximately a hundred yards from the Maryland Private Hospital for Ladies and Gentlemen he saw Doctor Keene, the family physician, descending the front steps, rubbing his hands together with a washing movement—as all doctors are required to do by the unwritten ethics of their profession.

Mr. Roger Button, the president of Roger Button & Co., Wholesale Hardware, began to run toward Doctor Keene with much less dignity than was expected from a Southern gentleman of that picturesque period. "Doctor Keene!" he called. "Oh, Doctor Keene!"

The doctor heard him, faced around, and stood waiting, a curious expression settling on his harsh, medicinal face as Mr. Button drew near.

"What happened?" demanded Mr. Button, as he came up in a gasping rush. "What was it? How is she? A boy? Who is it? What—"

"Talk sense!" said Doctor Keene sharply. He appeared somewhat irritated.

"Is the child born?" begged Mr. Button.

Doctor Keene frowned. "Why, yes, I suppose so—after a fashion." Again he threw a curious glance at Mr. Button.

Excerpt from The Curious Case of Benjamin Button by F.S. Fitzgerald, 1922

26. What major event is about to happen in this story?
    e. Mr. Button is about to go to a funeral.
    f. Mr. Button's wife is about to have a baby.
    g. Mr. Button is getting ready to go to the doctor's office.
    h. Mr. Button is about to go shopping for new clothes.

27. What kind of tone does the above passage have?
    a. Nervous and Excited
    b. Sad and Angry
    c. Shameful and Confused
    d. Grateful and Joyous

28. What is the meaning of the word "consecrated" in paragraph 4?
    e. Numbed
    f. Chained
    g. Dedicated
    h. Moved

29. What does the author mean to do by adding the following statement?

    "rubbing his hands together with a washing movement—as all doctors are required to do by the unwritten ethics of their profession."

    a. Suggesting that Mr. Button is tired of the doctor.
    b. Trying to explain the detail of the doctor's profession.
    c. Hinting to readers that the doctor is an unethical man.
    d. Giving readers a visual picture of what the doctor is doing.

30. Which of the following best describes the development of this passage?
    e. It starts in the middle of a narrative in order to transition smoothly to a conclusion.
    f. It is a chronological narrative from beginning to end.
    g. The sequence of events is backwards—we go from future events to past events.
    h. To introduce the setting of the story and its characters.

31. Which of the following is an example of an imperative sentence?
    a. "Oh, Doctor Keene!"
    b. "Talk sense!"
    c. "Is the child born?"
    d. "Why, yes, I suppose so—"

Questions 32–37 are based on the following passage:

Knowing that Mrs. Mallard was afflicted with heart trouble, great care was taken to break to her as gently as possible the news of her husband's death.

It was her sister Josephine who told her, in broken sentences; veiled hints that revealed in half concealing. Her husband's friend Richards was there, too, near her. It was he who had been in the newspaper office when intelligence of the railroad disaster was received, with Brently Mallard's name leading the list of "killed." He had only taken the time to assure himself of its truth by a second telegram, and had hastened to forestall any less careful, less tender friend in bearing the sad message.

She did not hear the story as many women have heard the same, with a paralyzed inability to accept its significance. She wept at once, with sudden, wild abandonment, in her sister's arms. When the storm of grief had spent itself she went away to her room alone. She would have no one follow her.

There stood, facing the open window, a comfortable, roomy armchair. Into this she sank, pressed down by a physical exhaustion that haunted her body and seemed to reach into her soul.

She could see in the open square before her house the tops of trees that were all aquiver with the new spring life. The delicious breath of rain was in the air. In the street below a peddler was crying his wares. The notes of a distant song which some one was singing reached her faintly, and countless sparrows were twittering in the eaves.

There were patches of blue sky showing here and there through the clouds that had met and piled one above the other in the west facing her window.

She sat with her head thrown back upon the cushion of the chair, quite motionless, except when a sob came up into her throat and shook her, as a child who has cried itself to sleep continues to sob in its dreams.

She was young, with a fair, calm face, whose lines bespoke repression and even a certain strength. But now here was a dull stare in her eyes, whose gaze was fixed away off yonder on one of those patches of blue sky. It was not a glance of reflection, but rather indicated a suspension of intelligent thought.

There was something coming to her and she was waiting for it, fearfully. What was it? She did not know; it was too subtle and elusive to name. But she felt it, creeping out of the sky, reaching toward her through the sounds, the scents, and color that filled the air.

Now her bosom rose and fell tumultuously. She was beginning to recognize this thing that was approaching to possess her, and she was striving to beat it back with her will—as powerless as her two white slender hands would have been. When she abandoned herself a little whispered word escaped her slightly parted lips. She said it over and over under her breath: "free, free, free!" The vacant stare and the look of terror that had followed it went from her eyes. They stayed keen and bright. Her pulses beat fast, and the coursing blood warmed and relaxed every inch of her body.

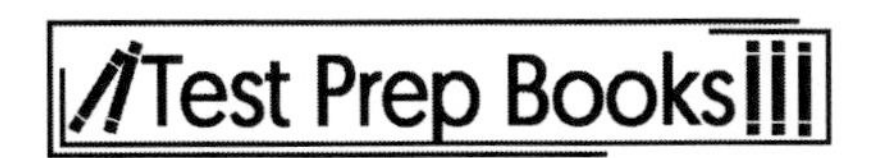

She did not stop to ask if it were or were not a monstrous joy that held her. A clear and exalted perception enabled her to dismiss the suggestion as trivial. She knew that she would weep again when she saw the kind, tender hands folded in death; the face that had never looked save with love upon her, fixed and gray and dead. But she saw beyond that bitter moment a long procession of years to come that would belong to her absolutely. And she opened and spread her arms out to them in welcome.

Excerpt from "The Story of An Hour" by Kate Chopin

32. What point of view is the above passage told in?
    e. First person
    f. Second person
    g. Third person omniscient
    h. Third person limited

33. What kind of irony are we presented with in this story?
    a. The way Mrs. Mallard reacted to her husband's death.
    b. The way in which Mr. Mallard died.
    c. The way in which the news of her husband's death was presented to Mrs. Mallard.
    d. The way in which nature is compared with death in the story.

34. What is the meaning of the word "elusive" in paragraph 9?
    e. Horrible
    f. Indefinable
    g. Quiet
    h. Joyful

35. What is the best summary of the passage above?
    a. Mr. Mallard, a soldier during World War I, is killed by the enemy and leaves his wife widowed.
    b. Mrs. Mallard understands the value of friendship when her friends show up for her after her husband's death.
    c. Mrs. Mallard combats mental illness daily and will perhaps be sent to a mental institution soon.
    d. Mrs. Mallard, a newly widowed woman, finds unexpected relief in her husband's death.

36. What is the tone of this story?
    e. Confused
    f. Joyful
    g. Depressive
    h. All of the above

37. What is the meaning of the word "tumultuously" in paragraph 10?
    a. Orderly
    b. Unashamedly
    c. Violently
    d. Calmly

Read the following poem and answer questions 38–44:

Four Seasons fill the measure of the year;
There are four seasons in the mind of man:
He has his lusty Spring, when fancy clear
Takes in all beauty with an easy span:
He has his Summer, when luxuriously *5*
Spring's honied cud of youthful thought he loves
To ruminate, and by such dreaming high
Is nearest unto heaven: quiet coves
His soul has in its Autumn, when his wings
He furleth close; contented so to look *10*
On mists in idleness—to let fair things
Pass by unheeded as a threshold brook.
He has his Winter too of pale misfeature,
Or else he would forego his mortal nature.

"The Human Seasons" by John Keats

38. What literary device does Keats primarily use in this poem?
   e. Simile
   f. Soliloquy
   g. Hyperbole
   h. Extended metaphor

39. The meaning of the word "ruminate" in line 7 is closest to:
   a. Ponder
   b. Unwind
   c. Respond
   d. Incorporate

40. According to the poem, how does a man change between Spring and Autumn?
   e. He starts preparing for his future.
   f. He feels more deeply connected to nature.
   g. He spends less time thinking about beautiful things.
   h. He becomes more sensible about how he spends his time.

41. Why does Keats end the poem with Winter?
   a. Winter represents the end of man's life.
   b. The narrator's least favorite season is winter.
   c. Winter is the final season of the calendar year.
   d. The poem is organized from the hottest season to the coldest.

42. Which statement would the narrator probably agree with?
   e. People are most content when they are young.
   f. People should appreciate the beauty of everyday life more.
   g. People change as they move through different stages of life.
   h. People spend too much time on daydreaming instead of being active.

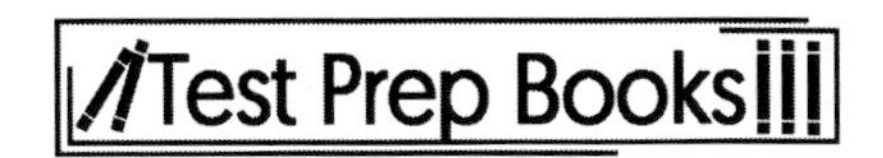

43. What does "he would forego his mortal nature" mean in the final line?
    a. He would take a break.
    b. He would postpone or avoid death.
    c. He would give up nature for technology.
    d. He would move away from the countryside.

44. Which of the following is an example of alliteration in this poem?
    e. "in the mind of man"
    f. "On mists of idleness"
    g. "his wings / He furleth closed"
    h. "unheeded as a threshold brook"

Read this article about NASA technology and answer questions 45–50:

> When researchers and engineers undertake a large-scale scientific project, they may end up making discoveries and developing technologies that have far wider uses than originally intended. This is especially true in NASA, one of the most influential and innovative scientific organizations in America. NASA *spinoff technology* refers to innovations originally developed for NASA space projects that are now used in a wide range of different commercial fields. Many consumers are unaware that products they are buying are based on NASA research! Spinoff technology proves that it's worthwhile to invest in science research because it could enrich people's lives in unexpected ways.
>
> The first spinoff technology worth mentioning is baby food. In space, where astronauts have limited access to fresh food and fewer options about their daily meals, malnutrition is a serious concern. Consequently, NASA researchers were looking for ways to enhance the nutritional value of astronauts' food. Scientists found that a certain type of algae could be added to food, improving the food's neurological benefits. When experts in the commercial food industry learned of this algae's potential to boost brain health, they were quick to begin their own research. The nutritional substance from algae then developed into a product called life's DHA, which can be found in over 90 percent of infant food sold in America.
>
> Another intriguing example of a spinoff technology can be found in fashion. People who are always dropping their sunglasses may have invested in a pair of sunglasses with scratch resistant lenses—that is, it's impossible to scratch the glass, even if the glasses are dropped on an abrasive surface. This innovation is incredibly advantageous for people who are clumsy, but most shoppers don't know that this technology was originally developed by NASA. Scientists first created scratch resistant glass to help protect costly and crucial equipment from getting scratched in space, especially the helmet visors in space suits. However, sunglasses companies later realized that this technology could be profitable for their products, and they licensed the technology from NASA.

45. What is the main purpose of this article?
    a. To advise consumers to do more research before making a purchase
    b. To persuade readers to support NASA research
    c. To tell a narrative about the history of space technology
    d. To define and describe examples of spinoff technology

46. What is the organizational structure of this article?
    e. A general definition followed by more specific examples
    f. A general opinion followed by supporting arguments
    g. An important moment in history followed by chronological details
    h. A popular misconception followed by counterevidence

47. Why did NASA scientists research algae?
    a. They already knew algae was healthy for babies.
    b. They were interested in how to grow food in space.
    c. They were looking for ways to add health benefits to food.
    d. They hoped to use it to protect expensive research equipment.

48. What does the word "neurological" mean in the second paragraph?
    e. Related to the body
    f. Related to the brain
    g. Related to vitamins
    h. Related to technology

49. Why does the author mention space suit helmets?
    a. To give an example of astronaut fashion
    b. To explain where sunglasses got their shape
    c. To explain how astronauts protect their eyes
    d. To give an example of valuable space equipment

50. Which statement would the author probably NOT agree with?
    e. Consumers don't always know the history of the products they are buying.
    f. Sometimes new innovations have unexpected applications.
    g. It's difficult to make money from scientific research.
    h. Space equipment is often very expensive.

Questions 51–56 are based on the following passage:

> People who argue that William Shakespeare isn't responsible for the plays attributed to his name are known as anti-Stratfordians (from the name of Shakespeare's birthplace, Stratford-upon-Avon). The most common anti-Stratfordian claim is that William Shakespeare simply was not educated enough or from a high enough social class to have written plays overflowing with references to such a wide range of subjects like history, the classics, religion, and international culture. William Shakespeare was the son of a glove-maker, he only had a basic grade school education, and he never set foot outside of England—so how could he have produced plays of such sophistication and imagination? How could he have written in such detail about historical figures and events, or about different cultures and locations around Europe? According to anti-Stratfordians, the depth of knowledge contained in Shakespeare's plays suggests a well-traveled writer from a wealthy background with a university education, not a countryside writer like Shakespeare. But in fact, there isn't much substance to such speculation, and most anti-Stratfordian arguments can be refuted with a little background about Shakespeare's time and upbringing.
>
> First of all, those who doubt Shakespeare's authorship often point to his common birth and brief education as stumbling blocks to his writerly genius. Although it's true that Shakespeare did not come from a noble class, his father was a very *successful* glove-maker and his mother was from

a very wealthy land-owning family—so while Shakespeare may have had a country upbringing, he was certainly from a well-off family and would have been educated accordingly. Also, even though he did not attend university, grade school education in Shakespeare's time was actually quite rigorous and exposed students to classic drama through writers like Seneca and Ovid. It's not unreasonable to believe that Shakespeare received a very solid foundation in poetry and literature from his early schooling.

Next, anti-Stratfordians tend to question how Shakespeare could write so extensively about countries and cultures he had never visited before (for example, several of his most famous works like *Romeo and Juliet* and *The Merchant of Venice* were set in Italy, on the opposite side of Europe). But again, this criticism doesn't hold up under scrutiny. For one thing, Shakespeare was living in London, a bustling metropolis of international trade, the most populous city in England, and a political and cultural hub of Europe. In the daily crowds of people, Shakespeare would certainly have been able to meet travelers from other countries and hear firsthand accounts of life in their home country. And, in addition to the influx of information from world travelers, this was also the age of the printing press, a jump in technology that made it possible to print and circulate books much more easily than in the past. This also allowed for a freer flow of information across different countries, allowing people to read about life and ideas throughout Europe. One needn't travel the continent in order to learn and write about its culture.

51. The main purpose of this article is to:
    a. Explain two sides of an argument and allow readers to choose which side they agree with
    b. Encourage readers to be skeptical about the authorship of famous poems and plays
    c. Give historical background about an important literary figure
    d. Criticize a theory by presenting counterevidence

52. Which sentence contains the author's thesis?
    e. "People who argue that William Shakespeare isn't responsible for the plays attributed to his name are known as anti-Stratfordians."
    f. "But in fact, there isn't much substance to such speculation, and most anti-Stratfordian arguments can be refuted with a little background about Shakespeare's time and upbringing."
    g. "It's not unreasonable to believe that Shakespeare received a very solid foundation in poetry and literature from his early schooling."
    h. "Next, anti-Stratfordians tend to question how Shakespeare could write so extensively about countries and cultures he had never visited before."

53. How does the author respond to the claim that Shakespeare was not well-educated because he didn't attend university?
    a. By insisting upon Shakespeare's natural genius
    b. By explaining grade school curriculum in Shakespeare's time
    c. By comparing Shakespeare with other uneducated writers of his time
    d. By pointing out that Shakespeare's wealthy parents probably paid for private tutors

54. What can be inferred from the article?
    e. Shakespeare's peers were jealous of his success and wanted to attack his reputation.
    f. Until recently, classical drama was only taught in universities.
    g. International travel was extremely rare in Shakespeare's time.
    h. In Shakespeare's time, glove-makers weren't part of the upper class.

55. Why does the author mention *Romeo and Juliet*?
    a. It's Shakespeare's most famous play.
    b. It was inspired by Shakespeare's trip to Italy.
    c. It's an example of a play set outside of England.
    d. It was unpopular when Shakespeare first wrote it.

56. Which statement would the author probably agree with?
    e. It's possible to learn things from reading rather than firsthand experience.
    f. If you want to be truly cultured, you need to travel the world.
    g. People never become successful without a university education.
    h. All of the world's great art comes from Italy.

Questions 57–62 are based on the following passages:

Passage I

Lethal force, or deadly force, is defined as the physical means to cause death or serious harm to another individual. The law holds that lethal force is only accepted when you or another person are in immediate and unavoidable danger of death or severe bodily harm. For example, a person could be beating a weaker person in such a way that they are suffering severe enough trauma that could result in death or serious harm. This would be an instance where lethal force would be acceptable and possibly the only way to save that person from irrevocable damage.

Another example of when to use lethal force would be when someone enters your home with a deadly weapon. The intruder's presence and possession of the weapon indicate mal-intent and the ability to inflict death or severe injury to you and your loved ones. Again, lethal force can be used in this situation. Lethal force can also be applied to prevent the harm of another individual. If a woman is being brutally assaulted and is unable to fend off an attacker, lethal force can be used to defend her as a last-ditch effort. If she is in immediate jeopardy of rape, harm, and/or death, lethal force could be the only response that could effectively deter the assailant.

The key to understanding the concept of lethal force is the term *last resort*. Deadly force cannot be taken back; it should be used only to prevent severe harm or death. The law does distinguish whether the means of one's self-defense is fully warranted, or if the individual goes out of control in the process. If you continually attack the assailant after they are rendered incapacitated, this would be causing unnecessary harm, and the law can bring charges against you. Likewise, if you kill an attacker unnecessarily after defending yourself, you can be charged with murder. This would move lethal force beyond necessary defense, making it no longer a last resort but rather a use of excessive force.

Passage II

Assault is the unlawful attempt of one person to apply apprehension on another individual by an imminent threat or by initiating offensive contact. Assaults can vary, encompassing physical strikes, threatening body language, and even provocative

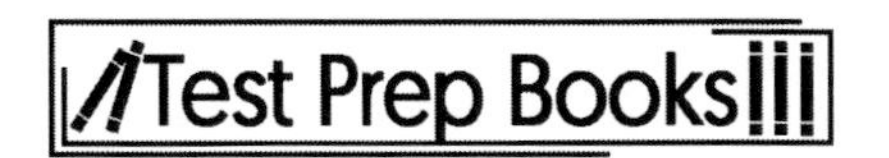

language. In the case of the latter, even if a hand has not been laid, it is still considered an assault because of its threatening nature.

Let's look at an example: A homeowner is angered because his neighbor blows fallen leaves into his freshly mowed lawn. Irate, the homeowner gestures a fist to his fellow neighbor and threatens to bash his head in for littering on his lawn. The homeowner's physical motions and verbal threat heralds a physical threat against the other neighbor. These factors classify the homeowner's reaction as an assault. If the angry neighbor hits the threatening homeowner in retaliation, that would constitute an assault as well because he physically hit the homeowner.

Assault also centers on the involvement of weapons in a conflict. If someone fires a gun at another person, this could be interpreted as an assault unless the shooter acted in self-defense. If an individual drew a gun or a knife on someone with the intent to harm them, that would be considered assault. However, it's also considered an assault if someone simply aimed a weapon, loaded or not, at another person in a threatening manner.

57. What is the purpose of the second passage?
   a. To inform the reader about what assault is and how it is committed
   b. To inform the reader about how assault is a minor example of lethal force
   c. To disprove the previous passage concerning lethal force
   d. The author is recounting an incident in which they were assaulted

58. In which of the following situations could lethal force be used, according to the passages, and not constitute an illegal use of lethal force?
   e. A disgruntled cash register yells obscenities at a customer.
   f. A thief is seen running away with stolen cash.
   g. A man is attacked in an alley by another man with a knife.
   h. A woman punches another woman in a bar.

59. Given the information in the passages, which of the following must be true about assault?
   a. Assault charges are more severe than unnecessary use of force charges.
   b. There are various forms of assault.
   c. Smaller, weaker people cannot commit assaults.
   d. Assault is justified only as a last resort.

60. Which of the following, if true, would most seriously undermine the explanation proposed by the author of Passage I in the third paragraph?
   e. An instance of lethal force in self-defense is not absolutely absolved from blame. The law considers the necessary use of force at the time it is committed.
   f. An individual who uses lethal force under necessary defense is in direct compliance of the law under most circumstances.
   g. Lethal force in self-defense should be forgiven in all cases for the peace of mind of the primary victim.
   h. The use of lethal force is not evaluated on the intent of the user, but rather the severity of the primary attack that warranted self-defense.

61. Based on the passages, what can be inferred about the relationship between assault and lethal force?
   a. An act of lethal force always leads to a type of assault.
   b. An assault will result in someone using lethal force.
   c. An assault with deadly intent can lead to an individual using lethal force to preserve their well-being.
   d. If someone uses self-defense in a conflict, it is called deadly force; if actions or threats are intended, it is called assault.

62. Which of the following best describes the way the passages are structured?
   e. Both passages open by defining a legal concept and then continue to describe situations that further explain the concept.
   f. Both passages begin with situations, introduce accepted definitions, and then cite legal ramifications.
   g. Passage I presents a long definition while the Passage II begins by showing an example of assault.
   h. Both cite specific legal doctrines, then proceed to explain the rulings.

Questions 63–67 are based on the following passage:

> In the quest to understand existence, modern philosophers must question if humans can fully comprehend the world. Classical western approaches to philosophy tend to hold that one can understand something, be it an event or object, by standing outside of the phenomena and observing it. It is then by unbiased observation that one can grasp the details of the world. This seems to hold true for many things. Scientists conduct experiments and record their findings, and thus many natural phenomena become comprehendible. However, several of these observations were possible because humans used tools in order to make these discoveries.
>
> This may seem like an extraneous matter. After all, people invented things like microscopes and telescopes in order to enhance their capacity to view cells or the movement of stars. While humans are still capable of seeing things, the question remains if human beings have the capacity to fully observe and see the world in order to understand it. It would not be an impossible stretch to argue that what humans see through a microscope is not the exact thing itself, but a human interpretation of it.
>
> This would seem to be the case in the "Business of the Holes" experiment conducted by Richard Feynman. To study the way electrons behave, Feynman set up a barrier with two holes and a plate. The plate was there to indicate how many times the electrons would pass through the hole(s). Rather than casually observe the electrons acting under normal circumstances, Feynman discovered that electrons behave in two totally different ways depending on whether or not they are observed. The electrons that were observed had passed through either one of the holes or were caught on the plate as particles. However, electrons that weren't observed acted as waves instead of particles and passed through both holes. This indicated that electrons have a dual nature. Electrons seen by the human eye act like particles, while unseen electrons act like waves of energy.
>
> This dual nature of the electrons presents a conundrum. While humans now have a better understanding of electrons, the fact remains that people cannot entirely perceive

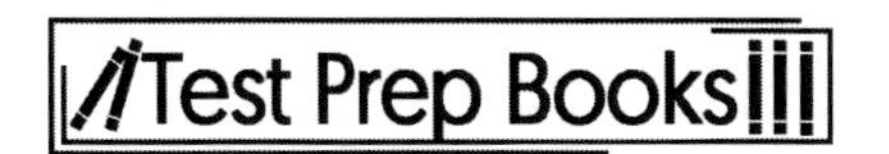

how electrons behave without the use of instruments. We can only observe one of the mentioned behaviors, which only provides a partial understanding of the entire function of electrons. Therefore, we're forced to ask ourselves whether the world we observe is objective or if it is subjectively perceived by humans. Or, an alternative question: can man understand the world only through machines that will allow them to observe natural phenomena?

Both questions humble man's capacity to grasp the world. However, those ideas don't consider that many phenomena have been proven by human beings without the use of machines, such as the discovery of gravity. Like all philosophical questions, whether man's reason and observation alone can understand the universe can be approached from many angles.

63. The word *extraneous* in paragraph two can be best interpreted as referring to which one of the following?
a. Indispensable
b. Bewildering
c. Superfluous
d. Exuberant

64. What is the author's motivation for writing the passage?
e. To bring to light an alternative view on human perception by examining the role of technology in human understanding.
f. To educate the reader on the latest astroparticle physics discovery and offer terms that may be unfamiliar to the reader.
g. To argue that humans are totally blind to the realities of the world by presenting an experiment that proves that electrons are not what they seem on the surface.
h. To reflect on opposing views of human understanding.

65. Which of the following most closely resembles the way in which paragraph four is structured?
a. It offers one solution, questions the solution, and then ends with an alternative solution.
b. It presents an inquiry, explains the details of that inquiry, and then offers a solution.
c. It presents a problem, explains the details of that problem, and then ends with more inquiry.
d. It gives a definition, offers an explanation, and then ends with an inquiry.

66. Which best describes how the electrons in the experiment behaved like waves?
e. The electrons moved up and down like actual waves.
f. The electrons passed through both holes and then onto the plate.
g. The electrons converted to photons upon touching the plate.
h. Electrons were seen passing through one hole or the other.

67. The author mentions "gravity" in the last paragraph in order to do what?
a. In order to show that different natural phenomena test man's ability to grasp the world.
b. To prove that since man has not measured it with the use of tools or machines, humans cannot know the true nature of gravity.
c. To demonstrate an example of natural phenomena humans discovered and understood without the use of tools or machines.
d. To show an alternative solution to the nature of electrons that humans have not thought of yet.

Math

1. If a car can travel 300 miles in 4 hours, how far can it go in an hour and a half?
   a. 100 miles
   b. 112.5 miles
   c. 135.5 miles
   d. 150 miles

2. At the store, Jan spends $90 on apples and oranges. Apples cost $1 each and oranges cost $2 each. If Jan buys the same number of apples as oranges, how many oranges did she buy?
   e. 20
   f. 25
   g. 30
   h. 35

3. What is the volume of a box with rectangular sides 5 feet long, 6 feet wide, and 3 feet high?
   a. 60 cubic feet
   b. 75 cubic feet
   c. 90 cubic feet
   d. 14 cubic feet

4. A train traveling 50 miles per hour takes a trip lasting 3 hours. If a map has a scale of 1 inch per 10 miles, how many inches apart are the train's starting point and ending point on the map?
   e. 14
   f. 12
   g. 13
   h. 15

5. A traveler takes an hour to drive to a museum, spends 3 hours and 30 minutes there, and takes half an hour to drive home. What percentage of his or her time was spent driving?
   a. 15%
   b. 30%
   c. 40%
   d. 60%

6. A truck is carrying three cylindrical barrels. Their bases have a diameter of 2 feet, and they have a height of 3 feet. What is the total volume of the three barrels in cubic feet?
   e. 3π
   f. 9π
   g. 12π
   h. 15π

7. Greg buys a $10 lunch with 5% sales tax. He leaves a $2 tip after his bill. How much money does he spend?
   a. $12.50
   b. $12
   c. $13
   d. $13.25

8. Which of the following is the result of simplifying the expression: $\frac{4a^{-1}b^3}{a^4b^{-2}} \times \frac{3a}{b}$?

e. $12a^3b^5$

f. $12\frac{b^4}{a^4}$

g. $\frac{12}{a^4}$

h. $7\frac{b^4}{a}$

9. What is 20% of 40?

a. 8
b.10
c.12
d.20

10. A couple buys a house for $150,000. They sell it for $165,000. By what percentage did the house's value increase?

e. 10%
f. 13%
g. 15%
h. 17%

11. A school has 15 teachers and 20 teaching assistants. They have 200 students. What is the ratio of faculty to students?

a. 3:20
b. 4:17
c. 5:54
d. 7:40

12. A map has a scale of 1 inch per 5 miles. A car can travel 60 miles per hour. If the distance from the start to the destination is 3 inches on the map, how long will it take the car to make the trip?

e. 12 minutes
f. 15 minutes
g. 17 minutes
h. 20 minutes

13. Taylor works two jobs. The first pays $20,000 per year. The second pays $10,000 per year. She donates 15% of her income to charity. How much does she donate each year?

a. $4500
b. $5000
c. $5500
d. $6000

14. Nathaniel has a box with the following dimensions. What is the surface area of the box?

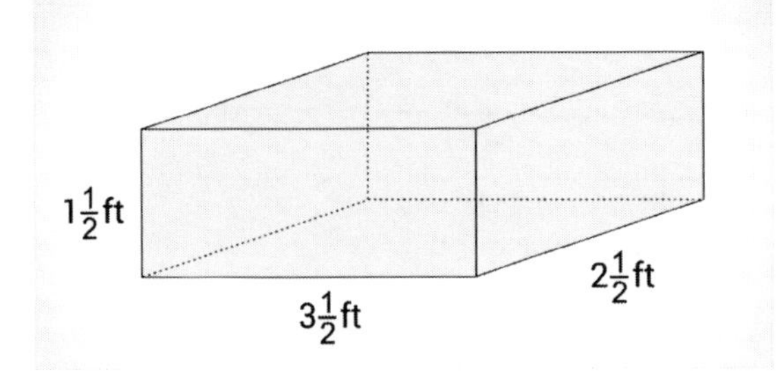

e. 13.125
f. 52.5
g. 35.5
h. 26.25

15. Kristen purchases \$100 worth of CDs and DVDs. The CDs cost \$10 each and the DVDs cost \$15. If she bought four DVDs, how many CDs did she buy?
a. 5
b. 6
c. 3
d. 4

16. If Sarah reads at an average rate of 21 pages in four nights, how long will it take her to read 140 pages?
e. 6 nights
f. 26 nights
g. 8 nights
h. 27 nights

17. Mom's car drove 72 miles in 90 minutes. There are 5280 feet per mile. How fast did she drive in feet per second?
a. 0.8 feet per second
b. 48.9 feet per second
c. 0.009 feet per second
d. 70.4 feet per second

18. This chart indicates how many sales of CDs, vinyl records, and MP3 downloads occurred over the last year. Approximately what percentage of the total sales was from CDs?

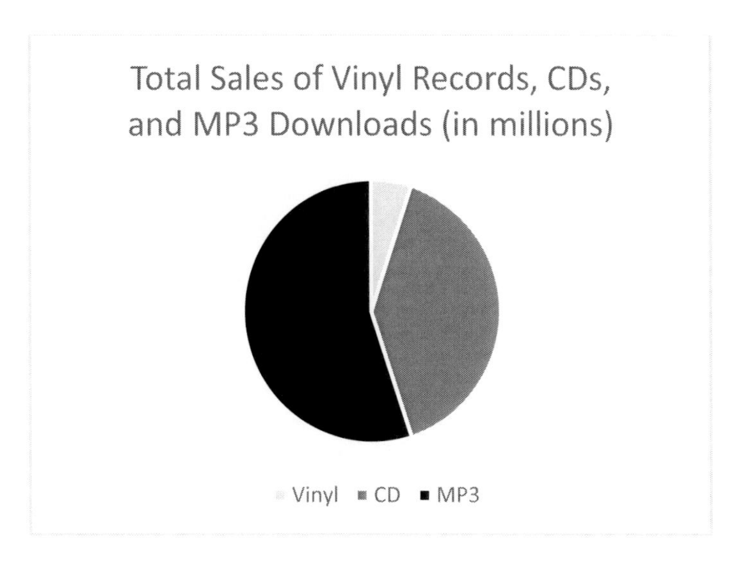

e. 55%
f. 25%
g. 40%
h. 5%

19. After a 20% sale discount, Frank purchased a new refrigerator for $850. How much did he save from the original price?
a. $170
b. $212.50
c. $105.75
d. $200

20. What is the simplified form of the expression $1.2 \times 10^{12} \div 3.0 \times 10^{8}$?
e. 0.4×10^{4}
f. 4.0×10^{4}
g. 4.0×10^{3}
h. 3.6×10^{20}

21. You measure the width of your door to be 36 inches. The true width of the door is 35.75 inches. What is the relative error in your measurement?
a. 0.7%
b. 0.007%
c. 0.99%
d. 0.1%

22. A ball is drawn at random from a ball pit containing 8 red balls, 7 yellow balls, 6 green balls, and 5 purple balls. What's the probability that the ball drawn is yellow?

e. $\frac{1}{26}$

f. $\frac{19}{26}$

g. $\frac{7}{26}$

h. 1

23. Two cards are drawn from a shuffled deck of 52 cards. What's the probability that both cards are Kings if the first card isn't replaced after it's drawn?

a. $\frac{1}{169}$

b. $\frac{1}{221}$

c. $\frac{1}{13}$

d. $\frac{4}{13}$

24. In the following figure, $\angle A$ is $2x$, and $\angle B$ is $4x - 6$. What is the measure of $\angle B$?

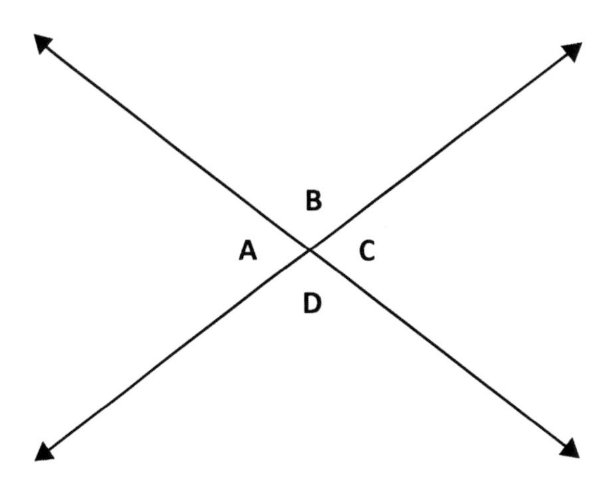

e. $31°$
f. $62°$
g. $118°$
h. $124°$

25. What's the probability of rolling a 6 at least once in two rolls of a die?

a. $\frac{1}{3}$

b. $\frac{1}{36}$

c. $\frac{1}{6}$

d. $\frac{11}{36}$

26. A student gets an 85% on a test with 20 questions. How many answers did the student solve correctly?

e. 15
f. 16
g. 17
h. 18

27. Four people split a bill. The first person pays for $\frac{1}{5}$, the second person pays for $\frac{1}{4}$, and the third person pays for $\frac{1}{3}$. What fraction of the bill does the fourth person pay?

a. $\frac{13}{60}$

b. $\frac{47}{60}$

c. $\frac{1}{4}$

d. $\frac{4}{15}$

28. 6 is 30% of what number?

e. 18
f. 20
g. 24
h. 26

29. $3\frac{2}{3} - 1\frac{4}{5} =$

a. $1\frac{13}{15}$

b. $\frac{14}{15}$

c. $2\frac{2}{3}$

d. $\frac{4}{5}$

30. For a group of 20 men, the median weight is 180 pounds, and the range is 30 pounds. If each man gains 10 pounds, which of the following would be true?
e. The median weight will increase, and the range will remain the same.
f. The median weight and range will both remain the same.
g. The median weight will stay the same, and the range will increase.
h. The median weight and range will both increase.

31. Dwayne has received the following scores on his math tests: 78, 92, 83, and 97. What score must Dwayne get on his next math test to have an overall average of at least 90?
a. 89
b. 98
c. 100
d. 94

32. Keith's bakery had 252 customers go through its doors last week. This week, that number increased to 378. By what percentage did his customer volume increase?
e. 26%
f. 50%
g. 35%
h. 12%

33. $52.3 \times 10^{-3} =$
a. 0.00523
b. 0.0523
c. 0.523
d. 523

34. If $\frac{5}{2} \div \frac{1}{3} = n$, then *n* is between:
e. 5 and 7
f. 7 and 9
g. 9 and 11
h. 3 and 5

35. Which inequality represents the following number line?

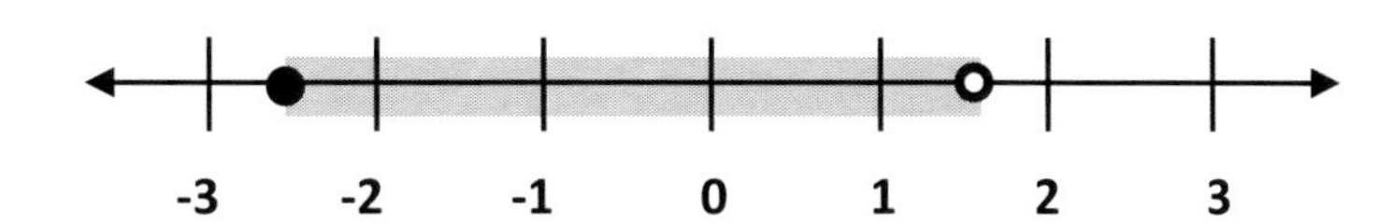

a. $-\frac{5}{2} \leq x < \frac{3}{2}$

b. $-\frac{7}{2} \leq x < \frac{5}{2}$

c. $-\frac{5}{2} <\leq \frac{3}{2}$

d. $\frac{5}{2} < x \leq -\frac{3}{2}$

36. Shawna buys $2\frac{1}{2}$ gallons of paint. If she uses $\frac{1}{3}$ of it on the first day, how much does she have left?

e. $1\frac{5}{6}$ gallons

f. $1\frac{1}{2}$ gallons

g. $1\frac{2}{3}$ gallons

h. 2 gallons

37. Which of the following inequalities is equivalent to $3 - \frac{1}{2}x \geq 2$?

a. $x \geq 2$
b. $x \leq 2$
c. $x \geq 1$
d. $x \leq 1$

38. What is the slope of this line?

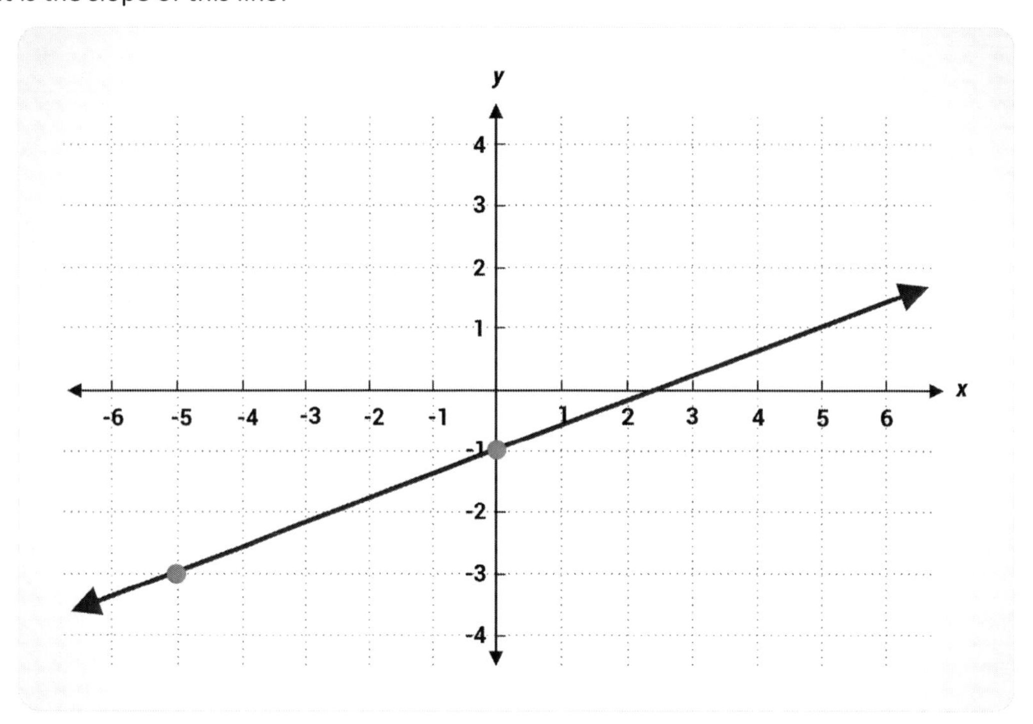

e. 2

f. $\frac{5}{2}$

g. $\frac{1}{2}$

h. $\frac{2}{5}$

39. What is the perimeter of the figure below? Note that the solid outer line is the perimeter.

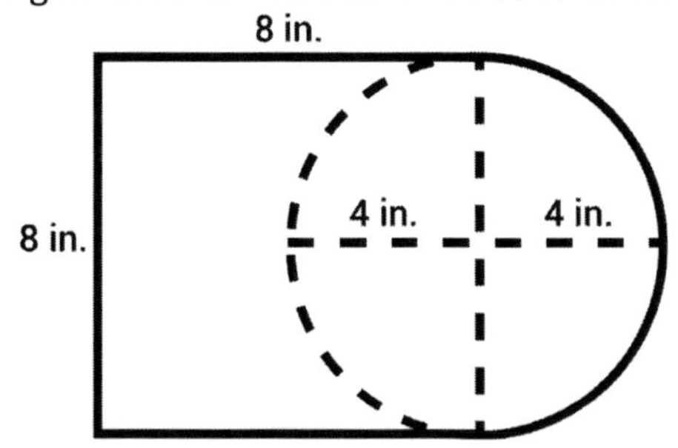

a. 48.565 in
b. 36.565 in
c. 39.78 in
d. 39.565 in

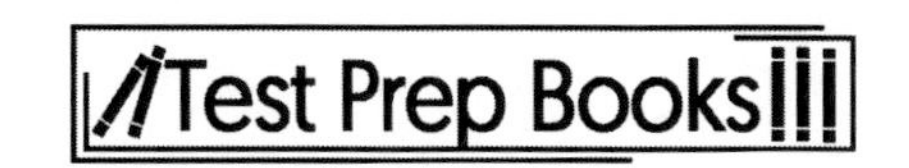

40. Which of the following equations best represents the problem below?
The width of a rectangle is 2 centimeters less than the length. If the perimeter of the rectangle is 44 centimeters, then what are the dimensions of the rectangle?

e. $2l + 2(l - 2) = 44$

f. $(l + 2) + (l + 2) + l = 48$

g. $l \times (l - 2) = 44$

h. $(l + 2) + (l + 2) + l = 44$

41. How will the following algebraic expression be simplified: $(5x^2 - 3x + 4) - (2x^2 - 7)$?

a. x^5
b. $3x^2 - 3x + 11$
c. $3x^2 - 3x - 3$
d. $x - 3$

42. In Jim's school, there are 3 girls for every 2 boys. There are 650 students in total. Using this information, how many students are girls?

e. 260
f. 130
g. 65
h. 390

43. Kimberley earns \$10 an hour babysitting, and after 10 p.m., she earns \$12 an hour, with the amount paid being rounded to the nearest hour accordingly. On her last job, she worked from 5:30 p.m. to 11 p.m. In total, how much did Kimberley earn for that job?

a. \$45
b. \$57
c. \$62
d. \$42

44. Five of six numbers have a sum of 25. The average of all six numbers is 6. What is the sixth number?

e. 8
f. 10
g. 11
h. 12

45. A local car dealership has compiled an inventory of cars on the lot by color. What percentage of cars on the lot are not black or white?

Color	Number of Cars
Black	56
White	48
Red	25
Gray	34
Blue	11
Tan	17

a. 54%
b. 46%
c. 52%
d. 43%

46. In May of 2010, a couple purchased a house for $100,000. In September of 2016, the couple sold the house for $93,000 so they could purchase a bigger one to start a family. How many months did they own the house?
e. 76
f. 54
g. 85
h. 93

47. At the beginning of the day, Xavier has 20 apples. At lunch, he meets his sister Emma and gives her half of his apples. After lunch, he stops by his neighbor Jim's house and gives him 6 of his apples. He then uses $\frac{3}{4}$ of his remaining apples to make an apple pie for dessert at dinner. At the end of the day, how many apples does Xavier have left?
a. 4
b. 6
c. 2
d. 1

48. What is the equation of a circle whose center is (0, 0) and whole radius is 5?
e. $(x-5)^2+(y-5)^2=25$
f. $(x)^2+(y)^2=5$
g. $(x)^2+(y)^2=25$
h. $(x+5)^2+(y+5)^2=25$

49. What is the solution to $4\times 7+(25-21)^2\div 2$?
a. 512
b. 36
c. 60.5
d. 22

50. The following graph compares the various test scores of the top three students in each of these teacher's classes. Based on the graph, which teacher's students had the smallest range of test scores?

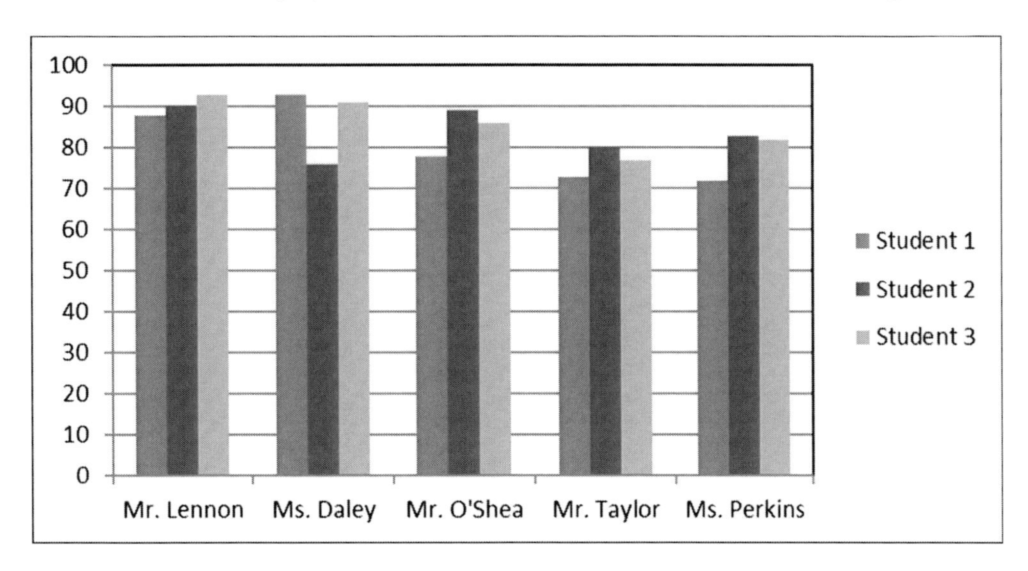

e. Mr. Lennon
f. Mr. O'Shea
g. Mr. Taylor
h. Ms. Daley

51. What is the volume of a cylinder, in terms of π, with a radius of 6 centimeters and a height of 2 centimeters?

a. 36π cm^3
b. 24π cm^3
c. 72π cm^3
d. 48π cm^3

52. What is the length of the hypotenuse of a right triangle with one leg equal to 3 centimeters and the other leg equal to 4 centimeters?

e. 7 cm
f. 5 cm
g. 25 cm
h. 12 cm

53. Arrange the following numbers from least to greatest value:

$0.85, \frac{4}{5}, \frac{2}{3}, \frac{91}{100}$

a. $0.85, \frac{4}{5}, \frac{2}{3}, \frac{91}{100}$

b. $\frac{4}{5}, 0.85, \frac{91}{100}, \frac{2}{3}$

c. $\frac{2}{3}, \frac{4}{5}, 0.85, \frac{91}{100}$

d. $0.85, \frac{91}{100}, \frac{4}{5}, \frac{2}{3}$

54. Simplify the following expression:

$$4\frac{2}{3} - 3\frac{4}{9}$$

e. $1\frac{1}{3}$

f. $1\frac{3}{8}$

g. 1

h. $1\frac{2}{9}$

55. Six people apply to work for Janice's company, but she only needs four workers. How many different groups of four employees can Janice choose?

a. 6
b. 10
c. 15
d. 36

56. Which of the following is equivalent to the value of the digit 3 in the number 792.134?

e. 3×10

f. 3×100

g. $\frac{3}{10}$

h. $\frac{3}{100}$

57. In the following expression, which operation should be completed first? $5 \times 6 + (5 + 4) \div 2 - 1$.

a. Multiplication
b. Addition
c. Division
d. Parentheses

58. How will the number 847.89632 be written if rounded to the nearest hundredth?
   e. 847.90
   f. 900
   g. 847.89
   h. 847.896

59. Which of the following is a mixed number?
   a. $16\frac{1}{2}$
   b. 16
   c. $\frac{16}{3}$
   d. $\frac{1}{4}$

60. Change 9.3 to a fraction.
   e. $9\frac{3}{7}$
   f. $\frac{903}{1000}$
   g. $\frac{9.03}{100}$
   h. $9\frac{3}{10}$

61. What is the value of *b* in this equation?

$$5b - 4 = 2b + 17$$

   a. 13
   b. 24
   c. 7
   d. 21

62. Katie works at a clothing company and sold 192 shirts over the weekend. One third of the shirts that were sold were patterned, and the rest were solid. Which mathematical expression would calculate the number of solid shirts Katie sold over the weekend?

   e. $192 \times \frac{1}{3}$
   f. $192 \div \frac{1}{3}$
   g. $192 \times (1 - \frac{1}{3})$
   h. $192 \div 3$

63. If Danny takes 48 minutes to walk 3 miles, how many minutes should it take him to walk 5 miles maintaining the same speed?

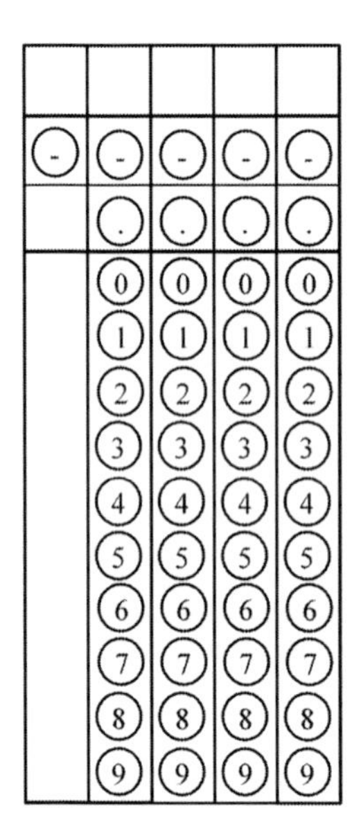

64. The perimeter of a 6-sided polygon is 56 cm. The length of three sides is 9 cm each. The length of two other sides is 8 cm each. What is the length of the missing side?

-	-	-	-	-
	.	.	.	.
	0	0	0	0
	1	1	1	1
	2	2	2	2
	3	3	3	3
	4	4	4	4
	5	5	5	5
	6	6	6	6
	7	7	7	7
	8	8	8	8
	9	9	9	9

65. Convert $\frac{3}{25}$ to a decimal.

-	-	-	-	-
	.	.	.	.
	0	0	0	0
	1	1	1	1
	2	2	2	2
	3	3	3	3
	4	4	4	4
	5	5	5	5
	6	6	6	6
	7	7	7	7
	8	8	8	8
	9	9	9	9

66. What is the value of $x^2 - 2xy + 2y^2$ when $x = 2, y = 3$?

-	-	-	-	-
	.	.	.	.
	0	0	0	0
	1	1	1	1
	2	2	2	2
	3	3	3	3
	4	4	4	4
	5	5	5	5
	6	6	6	6
	7	7	7	7
	8	8	8	8
	9	9	9	9

67. If $4x - 3 = 5$, what is the value of x?

Answer Explanations for Practice Test #3

Editing/Revising

1. B: Choice *B* is the correct answer. Commas are only needed before the word *and* when it serves as a conjunction combining two complete sentences. In sentence 2, there are not two complete sentences being joined together, so no comma is needed after the word *ocean*.

2. E: Choice *E* is the correct answer. Book titles such as *The Outsiders* should be underlined or italicized. A comma is not needed before the prepositional phrase *in 1967*.

3. D: Choice *D* is the correct answer. Sentence 4 is a complex sentence, meaning it consists of one independent clause and one dependent clause. *While collecting such a large number of cans would be impossible for one person* is a dependent clause, so a comma is needed after the word *person*. The remainder of the sentence, *collecting them around your city might be an idea worth entertaining*, is an independent clause.

4. H: Choice *H* combines the two sentences correctly, including all of the information and clarifying the relationship between the idea that back-to-school shopping can be very costly and the idea that Texas hosts a tax-free weekend to help parents with the expense. Choices *E, F,* and *G* are incorrect because each sentence focuses on parents with multiple children, which is not the main idea of the first sentence.

5. B: Choice *B* is the correct answer. The verb in this sentence should be past tense because the events in the sentence happened previously. The word *comes* should be *came,* and the word *is* should be *was*.

6. G: Choice *G* is the correct answer. This sentence combines all of the information provided into one cohesive sentence.

7. D: Choice *D* is the correct answer. In this sentence, the word *some* indicates more than one, so the singular word *instance* needs to be *instances*. The plural form of *injury* is spelled correctly by replacing the *y* with *ies* (*injuries*).

8. H: Choice *H* is the correct answer. The word *because* is a subordinating conjunction that introduces the phrase, *Because giraffes eat over 100 pounds of twigs and leaves a day.* This dependent clause needs to have a comma after the word *day*.

9. C: Choice *C* is correct, changing *Mythologies* to *mythology's*. Since one myth system is being referred to—and one particular component of it—the possessive is needed. Additionally, *Mythology's* does not need to be capitalized, since only the culture represents a proper noun. Choice *A* therefore is incorrect, with Choice *B* failing to fix the plural and Choice *D* having extraneous capitalization.

10. E: Choice *E* is correct because the sentence has no issues. While Choice *F* separates the sentence correctly, it makes more sense in this context of a direct comparison to keep the sentence intact. Choice *G* is incorrect because the sentence needs a comma after *superhero*. Choice *H* is unnecessarily long and lacks the word *but* that helps the author differentiate ideas.

11. B: Choice *B* is correct because it adds the two commas needed to clarify key subjects individually and establish a better flow to the sentence. Since *destroying mountains*, *defeating scores of giants*, and *lifting up the world's largest creature* are separate feats, commas are needed to separate them. Also, because *the world's largest creature* can stand alone in the sentence, a comma needs to proceed its name; *the Midgard Serpent* is not necessary to the sentence but rather provides extra information as an aside. Choice *A* is unclear and thus incorrect. Choice *C* is still missing a comma, while Choice *D* put an extraneous one in an incorrect place.

12. H: Choice *H* is correct since the sentence is lengthy as originally presented and should be split into two. Additionally, *however*, being a conjunction, needs a comma afterwards. Choice *E* is therefore incorrect due to missing punctuation. Choice *F* is an improvement but could separate the sentence's ideas better and more clearly. Choice *G* lacks the necessary comma after *However*.

13. A: Choice *A* is correct because this sentence has no issues with punctuation, content, or sentence construction. While there are three commas used, they serve to appropriately introduce an idea, an individual person, and transition into another line of thinking. Choices *B* and *C* miss commas needed to offset Loki's title as *the god of mischief*, while Choice *D* misses the comma needed to introduce the example.

14. G: Choice *G* is correct because the sentence needs two commas to emphasize the proper name of Mjolnir. Since Mjolnir is being talked about, directly addressed, and then explained, it must be flanked by commas to signify its role in the sentence. Choice *E* lacks necessary punctuation and is confusing. Choices *F* and *H* miss commas on either side of *Mjolnir*.

15. A: Choice *A* is correct, as this is an example of a compound sentence written correctly. Because of the conjunction *but* and the proceeding comma, the two independent clauses are able to form a single sentence coherently. While Choice *B* makes the question more direct, it doesn't go well with the remainder of the sentence. Choice *C* applies a comma after *but*, which is incorrect and confusing. Choice *D* inserts *however*, which is out of place and makes the sentence awkward.

16. F: Choice *F* correctly changes *an* to *a*, since *an* is only required when *a* precedes a word that begins with a vowel. Choice *E* therefore uses the incorrect form of *a*. Choice *G* fixes the issue but unnecessarily reverses the structure of the sentence, making it less direct and more confusing. Choice *H* does not fix the error and adds extraneous commas.

17. D: Choice *D* is correct, simply applying a comma after *Well* to introduce an idea. Choice *A* is therefore incorrect. Choice *B* introduces too many commas, resulting in a fractured sentence structure. Choice *C* applies a comma after *Well*, which is correct, but interrupts the flow of the sentence by switching the structure of the sentence. This makes the sentence lack fluidity and serves to confuse the reader.

18. G: Choice *G* is simple and straightforward, describing the event clearly for the reader to follow; talked is past tense, which is consistent with the rest of the passage. Choice *E* is incorrect, since we came to talking about confuses the tense of the sentence and the verb talk. Choices *F* and *H* are wordy and not as straightforward as Choice *G*.

19. B: Choice *B* is the correct answer because it adds a comma after *Essentially* and changes *an* to *a*. This is called the indefinite article, when an unspecified thing or quantity is referred to. However, *an* doesn't agree with *metaphor*, since *an* should only be used when the next word starts with a vowel. Choice *A* uses the article *an* and lacks the crucial comma after *Essentially*. Choice *C* is incorrect because it only provides the comma after *Essentially*, neglecting the indefinite article disagreement. Choice *D* is incorrect because neither issue is fixed and an unnecessary comma is introduced.

20. E: Choice *E* is correct because there are no errors present in the sentence. Choice *F* is a run on, because the clauses are not broken up by commas. Choice *G* has a verbal disagreement: *look* and *saw* are different tenses. Choice *H* changes the structure of the sentence but fails to add a transition to make this correct.

21. D: Choice *D* is correct because it uses a comma after the word *passage*, successfully connecting the dependent clause with its independent clause to form a complete thought/sentence. Choice *A* is therefore incorrect. Choice *B* uses a semicolon unwisely. The two clauses need to be connected to each other in order to make sense, otherwise they are just two fragments improperly combined. Choice *C* does not have the required comma and changes *found* to *finding*, an inappropriate tense for the verb in this sentence.

22. G: Choice *G* is correct because it fixes two major flaws in the original portion of the sentence. First, it inserts the adverb *also* to show the connection between *whatever it is that makes us unique* and *defines us*. Without this adverb, the sentence lacks clarity, and the connection is lost. Second, *big* is incorrect in this context. The sentence needs the superlative *bigger* in order to communicate the scope and scale of the author's assessment of how people relate to others on a grand scale. Choice *E* is therefore incorrect, Choice *F* inserts unnecessary commas, and Choice *H* subtly alters the original meaning.

23. B: Choice *B* is correct because a comma is correctly inserted after *ultimately*. This serves to express a side thought that helps transition into the rest of the sentence without having to break it apart. Choice *A* is incorrect because it lacks the comma after *ultimately*. Choice *C* uses too many commas and is overly complicated. Choice *D* lacks the necessary conjunction after the comma (*but)* before *ultimately*, making it a run-on sentence. It also lacks the important comma after *ultimately*.

24. H: Choice *H* corrects the subject-verb disagreement. *One's* is the possessive form of *one*, a single individual, not the plural *individuals. Isn't* is the singular contraction of *is not*, which conflicts with *individuals*. To correct this, either *isn't* must change to *aren't* or *individuals* should become the singular *individual*. The latter is correct because of the context of the sentence. Choice *E* is incorrect because of the subject-verb disagreement. Choice *F* uses the future tense, while Choice *G*'s *aren't* conflicts with *one's*, which is possessive singular.

25. A: Choice *A* contains no grammatical errors and communicates the writer's message clearly. Choice *B* inserts an unnecessary semicolon. Choice C uses *is*, which disagrees with the plural *aspects. Are* must be used because it is plural. This is the same for Choice *D*, which uses *there's* (*there is*).

Reading Comprehension

26. F: Mr. Button's wife is about to have a baby. The passage begins by giving the reader information about traditional birthing situations. Then, we are told that Mr. and Mrs. Button decide to go against tradition to have their baby in a hospital. The next few passages are dedicated to letting the reader know how Mr. Button dresses and goes to the hospital to welcome his new baby. There is a doctor in this excerpt, as Choice *G* indicates, and Mr. Button does put on clothes, as Choice *H* indicates. However, Mr. Button is not going to the doctor's office nor is he about to go shopping for new clothes.

27. A: The tone of the above passage is nervous and excited. We are told in the fourth paragraph that Mr. Button "arose nervously." We also see him running without caution to the doctor to find out about his wife and baby—this indicates his excitement. We also see him stuttering in a nervous yet excited fashion as he asks the doctor if it's a boy or girl. Though the doctor may seem a bit abrupt at the end, indicating a bit of anger or shame, neither of these choices is the overwhelming tone of the entire passage. Despite the circumstances, joy and gratitude are not the main tone in the passage.

28. G: Dedicated. Mr. Button is dedicated to the task before him. Choice *E*, numbed, Choice *F*, chained, and Choice *H*, moved, all could grammatically fit in the sentence. However, they are not synonyms with *consecrated* like Choice *G* is.

29. D: Giving readers a visual picture of what the doctor is doing. The author describes a visual image—the doctor rubbing his hands together—first and foremost. The author may be trying to make a comment about the profession; however, the author does not "explain the detail of the doctor's profession" as Choice *B* suggests.

30. H: To introduce the setting of the story and its characters. We know we are being introduced to the setting because we are given the year in the very first paragraph along with the season: "one day in the summer of 1860." This is a classic structure of an introduction of the setting. We are also getting a long explanation of Mr. Button, what his work is, who is related to him, and what his life is like in the third paragraph.

31. B: "Talk sense!" is an example of an imperative sentence. An imperative sentence gives a command. The doctor is commanding Mr. Button to talk sense. Choice *A* is an example of an exclamatory sentence, which expresses excitement. Choice *C* is an example of an interrogative sentence—these types of sentences ask questions. Choice *D* is an example of a declarative sentence. This means that the character is simply making a statement.

32. G: The point of view is told in third person omniscient. We know this because the story starts out with us knowing something that the character does not know: that her husband has died. Mrs. Mallard eventually comes to know this, but we as readers know this information before it is broken to her. In third person limited, Choice *H*, we would only see and know what Mrs. Mallard herself knew, and we would find out the news of her husband's death when she found out the news, not before.

33. A: The way Mrs. Mallard reacted to her husband's death. The irony in this story is called situational irony, which means the situation that takes place is different than what the audience anticipated. At the beginning of the story, we see Mrs. Mallard react with a burst of grief to her husband's death. However, once she's alone, she begins to contemplate her future and says the word "free" over and over. This is quite a different reaction from Mrs. Mallard than what readers expected from the first of the story.

34. F: The word "elusive" most closely means "indefinable." Horrible, Choice *E*, doesn't quite fit with the tone of the word "subtle" that comes before it. Choice *G*, "quiet," is more closely related to the word "subtle." Choice *H*, "joyful," also doesn't quite fit the context here. "Indefinable" is the best option.

35. D: Mrs. Mallard, a newly widowed woman, finds unexpected relief in her husband's death. A summary is a brief explanation of the main point of a story. The story mostly focuses on Mrs. Mallard and her reaction to her husband's death, especially in the room when she's alone and contemplating the present and future. All of the other answer choices except Choice *C* are briefly mentioned in the story; however, they are not the main focus of the story.

36. H: The interesting thing about this story is that feelings that are confused, joyful, and depressive all play a unique and almost equal part of this story. There is no one right answer here, because the author seems to display all of these emotions through the character of Mrs. Mallard. She displays feelings of depressiveness by her grief at the beginning; then, when she receives feelings of joy, she feels moments of confusion. We as readers cannot help but go through these feelings with the character. Thus, the author creates a tone of depression, joy, and confusion, all in one story.

37. C: The word "tumultuously" most nearly means "violently." Even if you don't know the word "tumultuously," look at the surrounding context to figure it out. The next few sentences we see Mrs. Mallard striving to "beat back" the "thing that was approaching to possess her." We see a fearful and almost violent reaction to the emotion that she's having. Thus, her chest would rise and fall tumultuously, or violently.

38. H: Extended metaphor. Metaphor is a direct comparison between two things, and extended metaphor is a lengthy, well-developed metaphor that usually extends over the length of the poem. In this poem, Keats forms an extended metaphor by drawing a comparison between the four seasons of nature and the "seasons" that humans experience from youth to old age.

39. A: Ponder. This question can be answered using context clues from the sentence: "Spring's honied cud of youthful thought he loves / To ruminate, and by such dreaming high / Is nearest unto heaven." Following the word "ruminate," it's restated as "such dreaming"; also, immediately before is the expression "youthful thought." Together, this sentence describes a young man pleasantly daydreaming. The only word related to thinking and daydreaming is "ponder," Choice *A*.

40. G: He spends less time thinking about beautiful things. This is a general comprehension question. The narrator describes a man in Autumn "contented so . . . to let fair things / Pass by unheeded." In this case, "fair" is another word for "beautiful," and letting things "pass by unheeded" means "he doesn't pay attention to them." In contrast, a man in the Spring and Summer of life spends time appreciating and daydreaming about beautiful things.

41. A: Winter represents the end of man's life. This is a purpose question, but it also requires readers to understand that this poem is an extended metaphor. Since the narrator is developing an extended comparison between seasons and life, it's natural that winter should come last because it's the season of death, dormancy, and "pale" nature (unlike, say, Spring, which is a season of life and rebirth in nature).

42. G: People change as they move through different stages of life. This is an inference question asking readers to understand the narrator's perspective. Choices *F* and *H* both include an opinion or advice to the reader, while the tone of the poem is more neutral or purely descriptive (the narrator is simply describing the stages of life, rather than advising readers on how to behave). Choice *G* more closely agrees with the comparison that the narrator sets up in the poem; just as seasons change in nature, people also change throughout their lives.

43. B: He would postpone or avoid death. This is both a vocabulary and a comprehension question. Based on the poem's extended metaphor, readers can gather that Winter is a metaphor for the end of life; all people must pass through Winter or else they would never die. Looking at the poem's vocabulary, "mortal" refers to human's limited life span (the opposite of "immortal"), and "forego" means to turn something down.

44. E: "in the mind of man" (2). This is a fairly straightforward question about literary devices. Alliteration refers to repetition of a word's beginning sound, and Choice *E* is the only example of that ("mind" and "man" both start with the letter M).

45. D: To define and describe examples of spinoff technology. This is a purpose question—*why* did the author write this? The article contains facts, definitions, and other objective information without telling a story or arguing an opinion. In this case, the purpose of the article is to inform the reader. The only answer choice related to giving information is Choice *D*: to define and describe.

46. E: A general definition followed by more specific examples. This organization question asks readers to analyze the structure of the essay. The topic of the essay is spinoff technology; the first paragraph gives a general definition of the concept, while the following two paragraphs offer more detailed examples to help illustrate this idea.

47. C: They were looking for ways to add health benefits to food. This reading comprehension question can be answered based on the second paragraph—scientists were concerned about astronauts' nutrition and began researching nutritional supplements. Choice *A* isn't true because it reverses the order of discovery (first NASA identified algae for astronaut use, and then it was further developed for use in baby food).

48. F: Related to the brain. This vocabulary question could be answered based on the reader's prior knowledge, but the passage provides context clues for readers who've never encountered the word "neurological." The next sentence talks about "this algae's potential to boost brain health," which is a paraphrase of "neurological benefits." From this context, readers should be able to infer that "neurological" relates to the brain.

49. D: To give an example of valuable space equipment. This purpose question requires readers to understand the relevance of the given detail. In this case, the author mentions "costly and crucial equipment" before space suit visors, which are given as an example of something valuable. Choice *A* isn't correct because fashion is only related to sunglasses, not to NASA equipment. Choice *B* can be eliminated because it's simply not mentioned. While Choice *C* seems like it could be true, it's not relevant.

50. G: It's difficult to make money from scientific research. The article gives several examples of how businesses have capitalized on NASA research, so it's unlikely that the author would agree with this statement. Evidence for the other answer choices can be found in the article: In Choice *E*, the author mentions that "many consumers are unaware that products they are buying are based on NASA research"; Choice *F* is a general definition of spinoff technology; and Choice *H* is mentioned in the final paragraph.

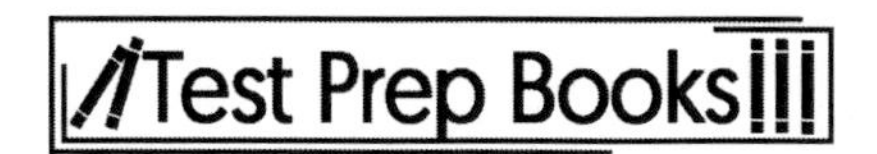

51. D: Criticize a theory by presenting counterevidence. The author mentions anti-Stratfordian arguments in the first paragraph, but then goes on to debunk these theories with facts about Shakespeare's life in the second and third paragraphs. Choice *A* is incorrect because the author is far from unbiased; in fact, the author clearly disagrees with anti-Stratfordians. Choice *B* is also incorrect because it's more closely aligned with the beliefs of anti-Stratfordians. Choice *C* can be eliminated because, while it's true that the author gives historical background, the purpose is using that information to disprove a theory.

52. F: "But in fact, there isn't much substance to such speculation, and most anti-Stratfordian arguments can be refuted with a little background about Shakespeare's time and upbringing." The thesis is a statement that contains the author's topic and main idea. As seen in the previous question, the purpose of this article is to use historical evidence to provide counterarguments to anti-Stratfordians. Choice *E* is simply a definition; Choice *G* is a supporting detail, not a main idea; and Choice *H* represents an idea of anti-Stratfordians, not the author's opinion.

53. B: By explaining grade school curriculum in Shakespeare's time. This question asks readers to refer to the organizational structure of the article and demonstrate understanding of how the author provides details to support the argument. This particular detail can be found in the second paragraph: "even though he did not attend university, grade school education in Shakespeare's time was actually quite rigorous."

54. H: In Shakespeare's time, glove-makers weren't part of the upper class. Anti-Stratfordians doubt Shakespeare's ability because he wasn't from the upper class; his father was a glove-maker; therefore, in at least this example, glove-makers weren't included in the upper class. This is an example of inductive reasoning, using two specific pieces of information to draw a more general conclusion.

55. C: It's an example of a play set outside of England. This detail comes from the third paragraph, where the author responds to skeptics who claim that Shakespeare wrote too much about places he never visited, so *Romeo and Juliet* is mentioned as a famous example of a play with a foreign setting. In order to answer this question, readers need to understand the author's purpose in the third paragraph and how the author uses details to support this purpose. Choices *A* and *D* aren't mentioned, and Choice *B* is clearly false because the passage mentions more than once that Shakespeare never left England.

56. E: It's possible to learn things from reading rather than firsthand experience. This inference can be made from the final paragraph, where the author refutes anti-Stratfordian skepticism by noting that books about life in Europe could circulate throughout London. From this statement, readers can conclude the author believes it's possible that Shakespeare learned about European culture from books. Choice *F* isn't true because the author believes that Shakespeare contributed to English literature without traveling extensively. Similarly, Choice *G* isn't a good answer because the author explains how Shakespeare got his education without attending a university. Choice *H* can also be eliminated because the author describes Shakespeare's genius, and Shakespeare clearly isn't from Italy.

57. A: The purpose is to inform the reader about what assault is and how it is committed. Choice *B* is incorrect because the passage does not state that assault is a lesser form of lethal force, only that an assault can use lethal force, or alternatively, lethal force can be utilized to counter a dangerous assault. Choice *C* is incorrect because the passage is informative and does not have a set agenda. Finally, Choice *D* is incorrect because although the author uses an example in order to explain assault, it is not indicated that this is the author's personal account.

58. G: If the man being attacked in an alley by another man with a knife used self-defense by lethal force, it would not be considered illegal. The presence of a deadly weapon indicates mal-intent and because the individual is isolated in an alley, lethal force in self-defense may be the only way to preserve his life. Choices *E* and *F* can be ruled out because in these situations, no one is in danger of immediate death or bodily harm by someone else. Choice *H* is an assault and does exhibit intent to harm, but this situation isn't severe enough to merit lethal force; there is no intent to kill.

59. B: As discussed in the second passage, there are several forms of assault, like assault with a deadly weapon, verbal assault, or threatening posture or language. Choice *A* is incorrect because the author does mention what the charges are on assaults; therefore, we cannot assume that they are more or less than unnecessary use of force charges. Choice *C* is incorrect because anyone is capable of assault; the author does not state that one group of people cannot commit assault. Choice *D* is incorrect because assault is never justified. Self-defense resulting in lethal force can be justified.

60. H: The use of lethal force is not evaluated on the intent of the user, but rather on the severity of the primary attack that warranted self-defense. This statement most undermines the last part of the passage because it directly contradicts how the law evaluates the use of lethal force. Choices *E* and *F* are stated in the paragraph, so they do not undermine the explanation from the author. Choice *G* does not necessarily undermine the passage, but it does not support the passage either. It is more of an opinion that does not offer strength or weakness to the explanation.

61. C: An assault with deadly intent can lead to an individual using lethal force to preserve their well-being. Choice *C* is correct because it clearly establishes what both assault and lethal force are and gives the specific way in which the two concepts meet. Choice *A* is incorrect because lethal force doesn't necessarily result in assault. This is also why Choice *B* is incorrect. Not all assaults would necessarily be life-threatening to the point where lethal force is needed for self-defense. Choice *D* is compelling but ultimately too vague; the statement touches on aspects of the two ideas but fails to present the concrete way in which the two are connected to each other.

62. E: Both passages open by defining a legal concept and then continue to describe situations in order to further explain the concept. Choice *H* is incorrect because while the passages utilize examples to help explain the concepts discussed, the author doesn't indicate that they are specific court cases. It's also clear that the passages don't open with examples, but instead, they begin by defining the terms addressed in each passage. This eliminates Choice *F,* and ultimately reveals Choice *E* to be the correct answer. Choice *E* accurately outlines the way both passages are structured. Because the passages follow a nearly identical structure, the Choice *G* can be ruled out.

63. C: *Extraneous* most nearly means *superfluous*, or *trivial*. Choice *A*, *indispensable*, is incorrect because it means the opposite of *extraneous*. Choice *B, bewildering*, means *confusing* and is not relevant to the context of the sentence. Finally, Choice *D* is incorrect because although the prefix of the word is the same, ex-, the word exuberant means elated or enthusiastic, and is irrelevant to the context of the sentence.

64. E: The author's purpose is to bring to light an alternative view on human perception by examining the role of technology in human understanding. This is a challenging question because the author's purpose is somewhat open-ended. The author concludes by stating that the questions regarding human perception and observation can be approached from many angles. Thus, the author does not seem to be attempting to prove one thing or another. Choice *F* is incorrect because we cannot know for certain whether the electron experiment is the latest discovery in astroparticle physics because no date is given.

Choice *G* is a broad generalization that does not reflect accurately on the writer's views. While the author does appear to reflect on opposing views of human understanding (Choice *H*), the best answer is Choice *E*.

65. C: It presents a problem, explains the details of that problem, and then ends with more inquiry. The beginning of this paragraph literally "presents a conundrum," explains the problem of partial understanding, and then ends with more questions, or inquiry. There is no solution offered in this paragraph, making Choices *A* and *B* incorrect. Choice *D* is incorrect because the paragraph does not begin with a definition.

66. F: The electrons passed through both holes and then onto the plate. Choices *E* and *G* are wrong because such movement is not mentioned at all in the text. In the passage, the author says that electrons that were physically observed appeared to pass through one hole or another. Remember, the electrons that were observed doing this were described as acting like particles. Therefore, Choice *H* is wrong. Recall that the plate actually recorded electrons passing through both holes simultaneously and hitting the plate. This behavior, the electron activity that wasn't seen by humans, was characteristic of waves. Thus, Choice *F* is the right answer.

67. C: The author uses "gravity" to demonstrate an example of natural phenomena humans discovered and understood without the use of tools or machines. Choice *A* mirrors the language in the beginning of the paragraph but is incorrect in its intent. Choice *B* is incorrect; the paragraph mentions nothing of "not knowing the true nature of gravity." Choice *D* is incorrect as well. There is no mention of an "alternative solution" to new technology in this paragraph.

Math

1. B: 300 miles in 4 hours is $300 \div 4 = 75$ miles per hour. In 1.5 hours, the car will go 1.5×75 miles, or 112.5 miles.

2. G: One apple/orange pair costs \$3 total. Therefore, Jan bought $90 \div 3 = 30$ total pairs, and hence, she bought 30 oranges.

3. C: The formula for the volume of a box with rectangular sides is the length times width times height, so:

$$5 \times 6 \times 3 = 90 \text{ cubic feet}$$

4. H: First, the train's journey in the real world is:

$$3 \times 50 = 150 \text{ miles}$$

On the map, 1 inch corresponds to 10 miles, so there is $150 \div 10 = 15$ inches on the map.

5. B: The total trip time is:

$$1 + 3.5 + 0.5 = 5 \text{ hours}$$

The total time driving is:

$$1 + 0.5 = 1.5 \text{ hours}$$

So, the fraction of time spent driving is:

$$\frac{1.5}{5}$$

or

$$\frac{3}{10}$$

To get the percentage, convert this to a fraction out of 100. The numerator and denominator are multiplied by 10, with a result of $\frac{30}{100}$. The percentage is the numerator in a fraction out of 100, so 30%.

6. F: The formula for the volume of a cylinder is $\pi r^2 h$, where *r* is the radius and *h* is the height. The diameter is twice the radius, so these barrels have a radius of 1 foot. That means each barrel has a volume of:

$$\pi \times 1^2 \times 3 = 3\pi \text{ cubic feet}$$

Since there are three of them, the total is $3 \times 3\pi = 9\pi$ cubic feet.

7. A: The tip is not taxed, so he pays 5% tax only on the \$10. 5% of \$10 is:

$$0.05 \times 10 = \$0.50$$

Add up $\$10 + \$2 + \$0.50$ to get $\$12.50$.

8. F: To simplify the given equation, the first step is to make all exponents positive by moving them to the opposite place in the fraction. This expression becomes:

$$\frac{4b^3b^2}{a^1a^4} \times \frac{3a}{b}$$

Then the rules for exponents can be used to simplify. Multiplying the same bases means the exponents can be added. Dividing the same bases means the exponents are subtracted. Thus, after multiplying the exponents in the first fraction the equation becomes:

$$\frac{4b^5}{a^5} \times \frac{3a}{b}$$

Therefore, we can first multiply to get $\frac{12ab^5}{a^5b}$. Then, dividing yields $12\frac{b^4}{a^4}$.

9. A: To find 20% of 40, simply multiply 40 by .20. This will give you the answer of 8.

10. E: The value went up by:

$$\$165{,}000 - \$150{,}000 = \$15{,}000$$

Out of $\$150{,}000$, this is:

$$\frac{15{,}000}{150{,}000} = \frac{1}{10}$$

Convert this to having a denominator of 100, the result is $\frac{10}{100}$ or 10%.

11. D: The total faculty is $15 + 20 = 35$. Therefore, the faculty to student ratio is 35:200. Then, to simplify this ratio, both the numerator and the denominator are divided by 5, since 5 is a common factor of both, which yields 7:40.

12. F: The journey will be $5 \times 3 = 15$ miles. A car traveling at 60 miles per hour is traveling at 1 mile per minute. The resulting equation would be:

$$\frac{15 \text{ mi}}{1\frac{\text{mi}}{\text{min}}} = 15 \text{ min}$$

Therefore, it will take 15 minutes to make the journey.

13. A: Taylor's total income is $\$20{,}000 + \$10{,}000 = \$30{,}000$. Fifteen percent of this is $\frac{15}{100} = \frac{3}{20}$. So:

$$\frac{3}{20} \times \$30{,}000 = \frac{\$90{,}000}{20}$$

$$\frac{\$9000}{2} = \$4500$$

14. G: The surface area formula for a rectangular prism is $2(lw + lh + wh)$. If the dimensions are converted to decimals, 3.5 ft can be used for length, 2.5 for width, and 1.5 for height. Substituting these values into the formula yields:

$$2(3.5 \times 2.5 + 3.5 \times 1.5 + 2.5 \times 1.5)$$

The surface area of the box is $35.5\ ft^2$.

15. D: Kristen bought four DVDs, which would cost a total of:

$$4 \times 15 = \$60$$

She spent a total of $100, so she spent:

$$\$100 - \$60 = \$40 \text{ on CDs}$$

Since they cost $10 each, she must have purchased:

$$40 \div 10 = 4 \text{ CDs}$$

16. H: This problem can be solved by setting up a proportion involving the given information and the unknown value. The proportion is:

$$\frac{21\ pages}{4\ nights} = \frac{140\ pages}{x\ nights}$$

Solving the proportion by cross-multiplying, the equation becomes $21x = 4 \times 140$, where $x = 26.67$. Since it is not an exact number of nights, the answer is rounded up to 27 nights. Twenty-six nights would not give Sarah enough time.

17. D: This problem can be solved by using unit conversion. The initial units are miles per minute. The final units need to be feet per second. Converting miles to feet uses the equivalence statement 1 mile equals 5,280 feet. Converting minutes to seconds uses the equivalence statement 1 minute equals 60 seconds. Setting up the ratios to convert the units is shown in the following equation:

$$\frac{72 \text{ mi}}{90 \text{ min}} \times \frac{1 \text{ min}}{60 \text{ s}} \times \frac{5280 \text{ ft}}{1 \text{ mi}} = 70.4 \frac{\text{ft}}{\text{s}}$$

The initial units cancel out, and the new units are left.

18. G: The sum total percentage of a pie chart must equal 100%. Since the CD sales take up less than half of the chart and more than a quarter (25%), it can be determined to be 40% overall. This can also be measured with a protractor. The angle of a circle is 360°. Since 25% of 360 would be 90° and 50% would be 180°, the angle percentage of CD sales falls in between; therefore, it would be Choice *G*.

19. B: Since $850 is the price *after* a 20% discount, $850 represents 80% of the original price. To determine the original price, set up a proportion with the ratio of the sale price (850) to original price (unknown) equal to the ratio of sale percentage (where x represents the unknown original price):

$$\frac{850}{x} = \frac{80}{100}$$

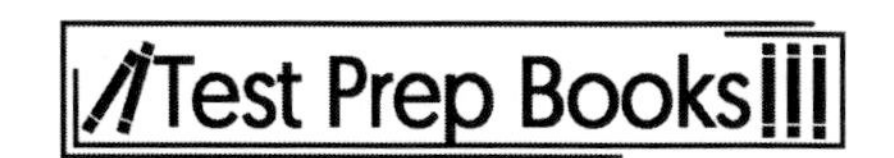

To solve a proportion, cross multiply the numerators and denominators and set the products equal to each other:

$$(850)(100) = (80)(x)$$

Multiplying each side results in the equation $85{,}000 = 80x$.

To solve for x, divide both sides by 80: $\frac{85{,}000}{80} = \frac{80x}{80}$, resulting in $x = 1062.5$. Remember that x represents the original price. Subtracting the sale price from the original price ($\$1062.50 - \$850$) indicates that Frank saved $212.50.

20. G: Scientific notation division can be solved by grouping the first terms together and grouping the tens together. The first terms can be divided, and the tens terms can be simplified using the rules for exponents. The initial expression becomes 0.4×10^4. This is not in scientific notation because the first number is not between 1 and 10. Shifting the decimal and subtracting one from the exponent, the answer becomes 4.0×10^3.

21. A: The relative error can be found by finding the absolute error and making it a percent of the true value. The absolute error is $36 - 35.75 = 0.25$. This error is then divided by 35.75—the true value—to find 0.7%.

22. G: The sample space is made up of:

$$8 + 7 + 6 + 5 = 26 \text{ balls}$$

The probability of pulling each individual ball is $\frac{1}{26}$. Since there are 7 yellow balls, the probability of pulling a yellow ball is $\frac{7}{26}$.

23. B: For the first card drawn, the probability of a King being pulled is $\frac{4}{52}$. Since this card isn't replaced, if a King is drawn first the probability of a King being drawn second is $\frac{3}{51}$. The probability of a King being drawn in both the first and second draw is the product of the two probabilities:

$$\frac{4}{52} \times \frac{3}{51} = \frac{12}{2652}$$

This fraction, when divided by 12, equals $\frac{1}{221}$.

24. G: $\angle A$ and $\angle B$ are supplementary angles, so together they equal $180°$. The value of x must be calculated first. The equation becomes:

$$2x + (4x - 6) = 180$$

Combining like terms yields $6x = 186$ so $x = 31$. Plugging the value of x into the equation for $\angle B$ results in:

$$4(31) - 6 = 118°$$

25. D: The addition rule is necessary to determine the probability because a 6 can be rolled on either roll of the die. The rule used is:

$$P(A \text{ or } B) = P(A) + P(B) - P(A \text{ and } B)$$

The probability of a 6 being individually rolled is $\frac{1}{6}$ and the probability of a 6 being rolled twice is:

$$\frac{1}{6} \times \frac{1}{6} = \frac{1}{36}$$

Therefore, the probability that a 6 is rolled at least once is:

$$\frac{1}{6} + \frac{1}{6} - \frac{1}{36} = \frac{11}{36}$$

26. G: 85% of a number means multiplying that number by 0.85. So:

$$0.85 \times 20 = \frac{85}{100} \times \frac{20}{1}$$

which can be simplified to:

$$\frac{17}{20} \times \frac{20}{1} = 17$$

27. A: To find the fraction of the bill that the first three people pay, the fractions need to be added, which means finding the common denominator. The common denominator will be 60.

$$\frac{1}{5} + \frac{1}{4} + \frac{1}{3} = \frac{12}{60} + \frac{15}{60} + \frac{20}{60} = \frac{47}{60}$$

The remainder of the bill is:

$$1 - \frac{47}{60} = \frac{60}{60} - \frac{47}{60} = \frac{13}{60}$$

28. F: 30% is $\frac{3}{10}$. The number itself must be $\frac{10}{3}$ of 6, or:

$$\frac{10}{3} \times 6 = 10 \times 2 = 20$$

29. A: Changing these numbers to improper fractions yields: $\frac{11}{3} - \frac{9}{5}$. Take 15 as a common denominator:

$$\frac{11}{3} - \frac{9}{5}$$

$$\frac{55}{15} - \frac{27}{15} = \frac{28}{15} = 1\frac{13}{15}$$

30. E: If each man gains 10 pounds, every original data point will increase by 10 pounds. Therefore, the man with the original median will still have the median value, but that value will increase by 10. The smallest value and largest value will also increase by 10 and, therefore, the difference between the two won't change. The range does not change in value and, thus, remains the same.

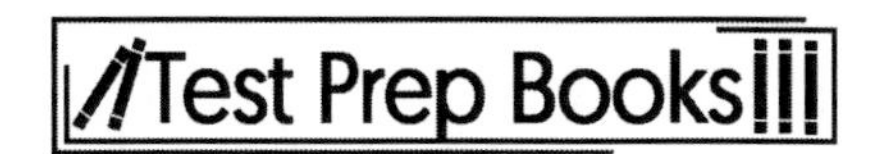

31. C: To find the average of a set of values, add the values together and then divide by the total number of values. In this case, include the unknown value of what Dwayne needs to score on his next test, in order to solve it.

$$\frac{78 + 92 + 83 + 97 + x}{5} = 90$$

Add the unknown value to the new average total, which is 5. Then multiply each side by 5 to simplify the equation, resulting in:

$$78 + 92 + 83 + 87 + x = 450$$

$$350 + x = 450$$

$$x = 100$$

Dwayne would need to get a perfect score of 100 in order to get an average of at least 90.

Test this answer by substituting back into the original formula.

$$\frac{78 + 92 + 83 + 97 + 100}{5} = 90$$

32. F: The first step is to calculate the difference between the larger value and the smaller value.

$$378 - 252 = 126$$

To calculate this difference as a percentage of the original value, and thus calculate the percentage *increase*, 126 is divided by 252, then this result is multiplied by 100 to find the percentage = 50%.

33. B: Multiplying by 10^{-3} means moving the decimal point three places to the left, putting in zeroes as necessary.

34. F: $\frac{5}{2} \div \frac{1}{3} = \frac{5}{2} \times \frac{3}{1} = \frac{15}{2} = 7.5$.

35. A: The solid dot is located between -2 and -3, and the open dot is located between 1 and 2. Therefore, x is between -2.5 and 1.5, which can be converted to $-\frac{5}{2}$ and $\frac{3}{2}$. The solid dot indicates greater than or equal to, and the open dot indicates less than so the inequality is:

$$-\frac{5}{2} \leq x < \frac{3}{2}$$

36. G: If she has used $\frac{1}{3}$ of the paint, she has $\frac{2}{3}$ remaining. $2\frac{1}{2}$ gallons are the same as $\frac{5}{2}$ gallons. The calculation is:

$$\frac{2}{3} \times \frac{5}{2} = \frac{5}{3} = 1\frac{2}{3} \text{ gallons}$$

37. B: To simplify this inequality, subtract 3 from both sides to get $-\frac{1}{2}x \geq -1$. Then, multiply both sides by -2 (remembering this flips the direction of the inequality) to get $x \leq 2$.

38. H: The slope is given by the change in *y* divided by the change in *x*. Specifically, it's:

$$slope = \frac{y_2 - y_1}{x_2 - x_1}$$

The first point is $(-5, -3)$ and the second point is $(0, \ -1)$. Work from left to right when identifying coordinates. Thus the point on the left is point 1 $(-5, \ -3)$, and the point on the right is point 2 $(0, \ -1)$.

Now we need to just plug those numbers into the equation:

$$lope = \frac{-1 - (-3)}{0 - (-5)}$$

It can be simplified to:

$$slope = \frac{-1 + 3}{0 + 5}$$

$$slope = \frac{2}{5}$$

39. B: The figure is composed of three sides of a square and a semicircle. The sides of the square are simply added:

$$8 + 8 + 8 = 24 \text{ inches}$$

The circumference of a circle is found by the equation $C = 2\pi r$. The radius is 4 in, so the circumference of the circle is 25.13 in. Only half of the circle makes up the outer border of the figure (part of the perimeter) so half of 25.13 in is 12.565 in. Therefore, the total perimeter is:

$$24\ in + 12.565\ in = 36.565\ in$$

The other answer choices use the incorrect formula or fail to include all of the necessary sides.

40. E: The first step is to determine the unknown, which is in terms of the length, l.

The second step is to translate the problem into the equation using the perimeter of a rectangle:

$$P = 2l + 2w$$

The width is the length minus 2 centimeters. The resulting equation is:

$$2l + 2(l - 2) = 44$$

The equation can be solved as follows:

$2l + 2l - 4 = 44$	Apply the distributive property on the left side of the equation
$4l - 4 = 44$	Combine like terms on the left side of the equation
$4l = 48$	Add 4 to both sides of the equation
$l = 12$	Divide both sides of the equation by 4

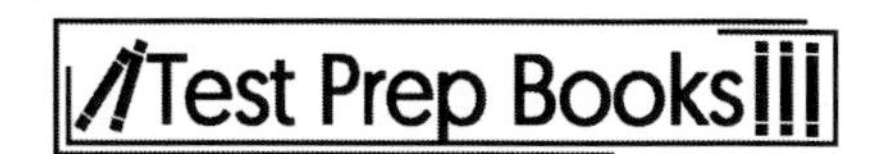

The length of the rectangle is 12 centimeters. The width is the length minus 2 centimeters, which is 10 centimeters. Checking the answers for length and width forms the following equation:

$$44 = 2(12) + 2(10)$$

The equation can be solved using the order of operations to form a true statement: $44 = 44$.

41. B: $3x^2 - 3x + 11$. By distributing the implied one in front of the first set of parentheses and the – 1 in front of the second set of parentheses, the parenthesis can be eliminated:

$$1(5x^2 - 3x + 4) - 1(2x^2 - 7)$$

$$5x^2 - 3x + 4 - 2x^2 + 7$$

Next, like terms (same variables with same exponents) are combined by adding the coefficients and keeping the variables and their powers the same:

$$5x^2 - 3x + 4 - 2x^2 + 7 = 3x^2 - 3x + 11$$

42. H: Three girls for every two boys can be expressed as a ratio: 3:2. This can be visualized as splitting the school into 5 groups: 3 girl groups and 2 boy groups. The number of students which are in each group can be found by dividing the total number of students by 5:

$$\frac{650 \text{ students}}{5 \text{ groups}} = \frac{130 \text{ students}}{\text{group}}$$

To find the total number of girls, multiply the number of students per group (130) by the number of girl groups in the school (3). This equals 390, Choice *H*.

43. C: Kimberley worked 4.5 hours at the rate of $10/h and 1 hour at the rate of $12/h. The problem states that her pay is rounded to the nearest hour, so the 4.5 hours would round up to 5 hours at the rate of $10/h.

$$5 \times \$10 + 1 \times \$12$$

$$\$50 + \$12 = \$62$$

44. G: The average is calculated by adding all six numbers, then dividing by 6. The first five numbers have a sum of 25. If the total divided by 6 is equal to 6, then the total itself must be 36. The sixth number must be $36 - 25 = 11$.

45. B: To find the percentage of cars that are not black or white, the total number of cars must be found first. The total number of cars is 191, and the total of all the cars that are not black or white is 87. The percentage can be found using the following calculation:

$$\frac{87}{191} = 0.455 = 46\%$$

46. E: This problem can be solved by simple multiplication and addition. Since the sale date is over six years apart, 6 can be multiplied by 12 for the number of months in a year, and then the remaining 4 months can be added.

$$(6 \times 12) + 4 = ?$$

$$72 + 4 = 76$$

47. D: This problem can be solved using basic arithmetic. Xavier starts with 20 apples, then gives his sister half, so 20 divided by 2.

$$\frac{20}{2} = 10$$

He then gives his neighbor 6, so 6 is subtracted from 10.

$$10 - 6 = 4$$

Lastly, he uses ¾ of his apples to make an apple pie, so to find remaining apples, the first step is to subtract ¾ from one and then multiply the difference by 4.

$$\left(1 - \frac{3}{4}\right) \times 4 = ?$$

$$\left(\frac{4}{4} - \frac{3}{4}\right) \times 4 = ?$$

$$\left(\frac{1}{4}\right) \times 4 = 1$$

48. G: Nothing is added to *x* and *y* since the center is 0 and 5^2 is 25. Choice *E* is not the correct answer because you do not subtract the radius from *x* and *y*. Choice *F* is not the correct answer because you must square the radius on the right side of the equation. Choice *H* is not the correct answer because you do not add the radius to *x* and *y* in the equation.

49. B: To solve this correctly, keep in mind the order of operations with the mnemonic PEMDAS (Please Excuse My Dear Aunt Sally). This stands for Parentheses, Exponents, Multiplication, Division, Addition, Subtraction. Taking it step by step, solve the parentheses first:

$$4 \times 7 + 4^2 \div 2$$

Then, apply the exponent:

$$4 \times 7 + 16 \div 2$$

Multiplication and division are both performed next:

$$28 + 8 = 36$$

Addition and subtraction are done last. The solution is 36.

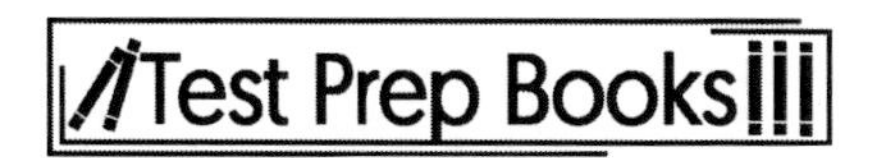

50. E: To calculate the range in a set of data, subtract the lowest value from the highest value. In this graph, the range of Mr. Lennon's students is 5, which can be seen physically in the graph as having the smallest difference between the highest value and the lowest value compared with the other teachers.

51. C: The volume of a cylinder is $\pi r^2 h$, and $\pi \times 6^2 \times 2$ is $72\,\pi$ cm^3. Choice *A* is not the correct answer because that is only $6^2 \times \pi$. Choice *B* is not the correct answer because that is $2^2 \times 6 \times \pi$. Choice *D* is not the correct answer because that is $2^3 \times 6 \times \pi$.

52. F: This answer is correct because $3^2 + 4^2$ is $9 + 16$, which is 25. Taking the square root of 25 is 5. Choice *E* is not the correct answer because that is $3 + 4$. Choice *G* is not the correct answer because that is stopping at $3^2 + 4^2$ is $9 + 16$, which is 25. Choice *H* is not the correct answer because that is 3×4.

53. C: The first step is to depict each number using decimals. $\frac{91}{100}$ = 0.91. Dividing the numerator by denominator of $\frac{4}{5}$ to convert it to a decimal yields 0.80, while $\frac{2}{3}$ becomes 0.66 recurring. Rearrange each expression in ascending order, as found in Choice *C*.

54. H: Simplify each mixed number of the problem into a fraction by multiplying the denominator by the whole number and adding the numerator:

$$\frac{14}{3} - \frac{31}{9}$$

Since the first denominator is a multiple of the second, simplify it further by multiplying both the numerator and denominator of the first expression by 3 so that the denominators of the fractions are equal:

$$\frac{42}{9} - \frac{31}{9} = \frac{11}{9}$$

Simplifying this further, divide the numerator 11 by the denominator 9; this leaves 1 with a remainder of 2. To write this as a mixed number, place the remainder over the denominator, resulting in $1\frac{2}{9}$.

55. C: Janice will be choosing 4 employees out of a set of 6 applicants, so this will be given by the choice function. The following equation shows the choice function worked out:

$$\binom{6}{4} = \frac{6!}{4!\,(6-4)!} = \frac{6!}{4!\,(2)!}$$

$$\frac{6 \times 5 \times 4 \times 3 \times 2 \times 1}{4 \times 3 \times 2 \times 1 \times 2 \times 1} = \frac{6 \times 5}{2} = 15$$

56. H: $\frac{3}{100}$. Each digit to the left of the decimal point represents a higher multiple of 10 and each digit to the right of the decimal point represents a quotient of a higher multiple of 10 for the divisor.

The first digit to the right of the decimal point is equal to the value $\div$ 10. The second digit to the right of the decimal point is equal to the value $\div$ (10×10), or the value $\div$ 100.

57. D: Using the order of operations, items inside of parentheses are sorted out first. Therefore, the teacher should resolve the parentheses first. In this expression, multiplication and division are computed next from left to right, followed by addition and subtraction.

58. E: 847.90.

The hundredths place value is located two digits to the right of the decimal point (the digit 9 in the original number). The digit to the right of the place value is examined to decide whether to round up or keep the digit. In this case, the digit 6 is 5 or greater so the hundredth place is rounded up. When rounding up, if the digit to be increased is a 9, the digit to its left is increased by one and the digit in the desired place value is made a zero. Therefore, the number is rounded to 847.90.

59. A: $16\frac{1}{2}$. A mixed number contains both a whole number and either a fraction or a decimal. Therefore, the mixed number is $16\frac{1}{2}$.

60. H: $9\frac{3}{10}$. To convert a decimal to a fraction, remember that any number to the left of the decimal point will be a whole number. Then, since 0.3 goes to the tenths place, it can be placed over 10.

61. C: To solve for the value of *b*, both sides of the equation need to be equalized.

Start by cancelling out the lower value of -4 by adding 4 to both sides:

$$5b - 4 = 2b + 17$$
$$5b - 4 + 4 = 2b + 17 + 4$$
$$5b = 2b + 21$$

The variable *b* is the same on each side, so subtract the lower 2b from each side:

$$5b = 2b + 21$$
$$5b - 2b = 2b + 21 - 2b$$
$$3b = 21$$

Then divide both sides by 3 to get the value of *b*:

$$3b = 21$$

$$\frac{3b}{3} = \frac{21}{3}$$

$$b = 7$$

62. G: $\frac{1}{3}$ of the shirts sold were patterned. Therefore, $1 - \frac{1}{3} = \frac{2}{3}$ of the shirts sold were solid. Anytime "of" a quantity appears in a word problem, multiplication needs to be used. Therefore:

$$192 \times \frac{2}{3} = 192 \times \frac{2}{3} = \frac{384}{3} = 128 \text{ solid shirts were sold}$$

The entire expression is:

$$192 \times \left(1 - \frac{1}{3}\right)$$

			8	0

63. 80. To solve the problem, a proportion is written consisting of ratios comparing distance and time. One way to set up the proportion is:

$$\frac{3}{48} = \frac{5}{x} \left(\frac{distance}{time} = \frac{distance}{time}\right)$$

where x represents the unknown value of time. To solve a proportion, the ratios are cross-multiplied:

$$(3)(x) = (5)(48) \rightarrow 3x = 240$$

The equation is solved by isolating the variable, or dividing by 3 on both sides, to produce $x = 80$.

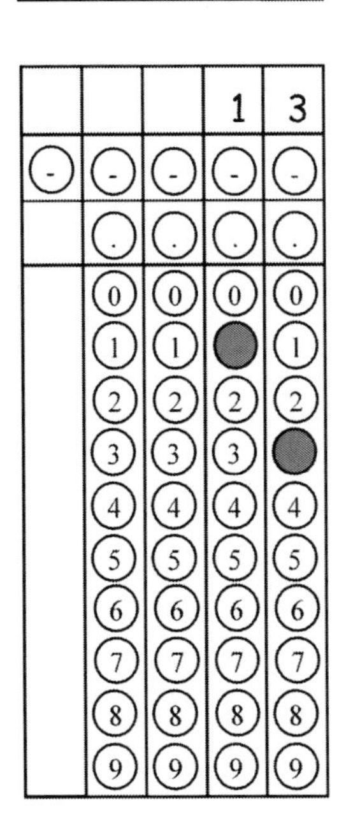

64. 13. Perimeter is found by calculating the sum of all sides of the polygon. $9 + 9 + 9 + 8 + 8 + s = 56$, where s is the missing side length. Therefore, 43 plus the missing side length is equal to 56. The missing side length is 13 cm.

	0	.	1	2

65. 0.12. The fraction is converted so that the denominator is 100 by multiplying the numerator and denominator by 4, to get $\frac{3}{25} = \frac{12}{100}$. Dividing a number by 100 just moves the decimal point two places to the left, with a result of 0.12.

			1	0

66. 10. Start with the original equation: $x^2 - 2xy + 2y$, then replace each instance of x with a 2, and each instance of y with a 3 to get:

$$2^2 - 2 \times 2 \times 3 + 2 \times 3^2 = 4 - 12 + 18 = 10$$

67. 2. Add 3 to both sides to get $4x = 8$. Then divide both sides by 4 to get:

$$x = 2.$$

Dear SHSAT Test Taker,

We would like to start by thanking you for purchasing this study guide for your SHSAT exam. We hope that we exceeded your expectations.

Our goal in creating this study guide was to cover all of the topics that you will see on the test. We also strove to make our practice questions as similar as possible to what you will encounter on test day. With that being said, if you found something that you feel was not up to your standards, please send us an email and let us know.

We would also like to let you know about other books in our catalog that may interest you.

SAT

amazon.com/dp/1628456868

ACT

amazon.com/dp/1628459468

ACCUPLACER

amazon.com/dp/1628459344

AP Biology

amazon.com/dp/1628456221

We have study guides in a wide variety of fields. If the one you are looking for isn't listed above, then try searching for it on Amazon or send us an email.

Thanks Again and Happy Testing!
Product Development Team
info@studyguideteam.com

FREE Test Taking Tips DVD Offer

To help us better serve you, we have developed a Test Taking Tips DVD that we would like to give you for FREE. **This DVD covers world-class test taking tips that you can use to be even more successful when you are taking your test.**

All that we ask is that you email us your feedback about your study guide. Please let us know what you thought about it – whether that is good, bad or indifferent.

To get your **FREE Test Taking Tips DVD**, email freedvd@studyguideteam.com with "FREE DVD" in the subject line and the following information in the body of the email:

a. The title of your study guide.

b. Your product rating on a scale of 1-5, with 5 being the highest rating.

c. Your feedback about the study guide. What did you think of it?

d. Your full name and shipping address to send your free DVD.

If you have any questions or concerns, please don't hesitate to contact us at freedvd@studyguideteam.com.

Thanks again!